PROMOTIONAL BLURB FROM FAMOUS PERSON | *No.* **3185**

"The smart founders of *ReadyMade* magazine offer up a sophisticated, clever, and raw approach to design, guaranteed to thrill and inspire modern do-it-yourselfers of any age."

—Todd Oldham

The surrealist artist Marcel Duchamp came up with the term *readymade* to describe any ordinary object—a snow shovel, a comb, a urinal—that he'd altered, signed, and deemed art. Duchamp is this book's patron saint.

"Anything can
be beautiful."

1. CATCH

3. FEATHER

HOW TO MAKE {ALMOST} EVERYTHING

A Do-It-Yourself Primer

SHOSHANA BERGER
TEXT

GRACE HAWTHORNE
PROJECTS

Jeffery Cross photography
Kate Francis illustrations
Eric Heiman/Volume design

Thames & Hudson

190
174
117
METAL
108
88
189
186
160
GLASS
138
54
148
Fabric
168
20
50
144
157
16
68
154
APPENDIX
198
GLOSSARY
200
CREDITS &
ACKNOWLEDGMENTS
202
INDEX
203

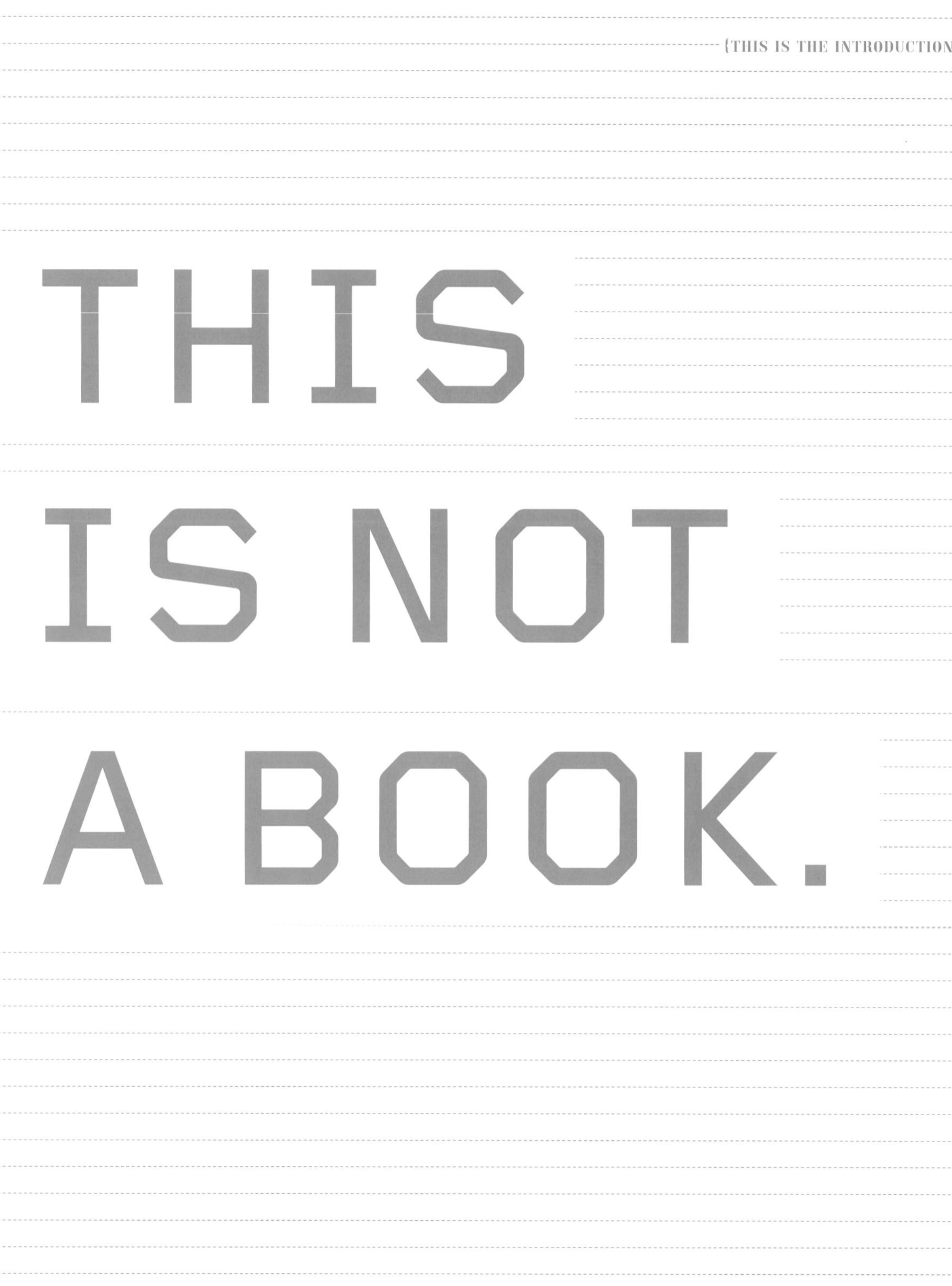
THIS
IS NOT
A BOOK.

IT IS A PICTURE FRAME,

a straight edge, a demonstration of how gravity works (look how swiftly it falls from your hands!), and a good anchor when your balloon is in danger of floating away.

Ignore its function as a physical object; it is an atlas of suggestions, if, as we do, you take suggestions to be little bolts of energy sent down from the universe to trigger a salvo in your brain. Perhaps you will try making a thing or two. Perhaps you only imagine the day when you will make something. Either way, we hope that the projects you see here have a catalytic effect, inspiring you to rethink the purpose of an old telephone book or bicycle wheel. The motley crew of projects in the pages ahead are not meant to discharge some grand, marquis-worthy event, a new Bauhaus or High Art movement. Likewise, we do not claim to be the first among all humans to have conceived of these ideas. (We made our best attempt to cite influences; listing them all would be like trying to count every grain of sand on a beach. We are the product of a densely visual culture, and, like Palmolive, we're soaking in it.) Nor are we trained designers or engineers. Indeed, we have no special skills whatsoever. What we do claim is a belief that we can make things work by some mash-up of imagination and sheer force of will. Like Marcel Duchamp, the twentieth-century surrealist artist from whom we borrowed the word *readymade*, we believe in artists more than art. All we do here is identify some ideal candidates for reuse, things so utilitarian and mundane—shopping bags, water bottles, chopsticks—that we hardly notice how much we depend on them just to get through the day.

Our tactical plan: Break a few of these items down into their constituent elements—paper, plastic, wood, metal, glass, and fabric; choose representatives from each category as the material for our projects; then make sure each project meets a rigorous set of criteria. To pass Go, the projects had to have at least four of the following qualities: look handsome, incorporate a common object, challenge the do-it-yourselfer (without being unnecessarily difficult or, gulp, dangerous), recontextualize the material at hand, and inspire by example. Of course, garage invention is no picnic. Setting out to make what you've dreamt up at the kitchen table is a little like trying to rediscover fire—you're lucky if you get a single spark on the first try. There were equal parts trial and error, magnificent failure and improbable success. Fearlessly, we'll show you both.

Beyond making stuff with the items we choose, we also dig a little deeper into the singular charms of each, investigating how plastic is made, how long people have used wood to make weapons, which metals are good for you to eat and which are not. We had many material science questions, and set out to answer them in the only way we know how: Google. You'll also find two essays per chapter that are loosely based on the materials at hand, as well as personal accounts of how a mundane thing like bubble wrap or a glass plate is conferred with new meaning when it becomes a central character in the stories of our lives. These tiny tales can be found at the bottom of the pages, in small print. We apologize in advance if reading them requires the use of a magnifying glass. We ran out of room. If it's any consolation, we're having vision problems, too. There's this gray floater that swims around in front of our eyes in the morning, like a spot that was burned out by looking too long at the sun.

Lastly, fun! Do not forget to have fun along the way. Think of each project as a pit stop in the creative process, not a final destination. We fully expect you to get behind the wheel, floor it, and go places we never knew existed.

A MINI MANIFESTO

{with apologies to William Blake}

I swear to consider (for several minutes) the following tenets issued by *ReadyMade:*

01 I will in some way redefine space, material, functionality, or context.

02 I understand that the phrase "home décor" ist verboten.

03 With each project, I will disclose the means of production, invite collaboration, and generally debunk the maker-as-auteur theory.

04 I acknowledge that common, everyday objects are precious gems.

05 I will attempt to keep all consumer goods in circulation, and out of the big Wal-Mart in the sky, by reusing them.

06 Excess of simplicity laughs, excess of complexity weeps!

07 All creative endeavors are a balancing act of Newtonian physics, kitchen-table logic, and snafu. We ask that you forgive what you do not approve and love us for this energetic exertion of our talent.

And. . . Furthermore, I swear as a *ReadyMade* reader to refrain from creating a "finished work," as a heightened awareness of process is the whole point. I will share this book with friend and foe alike with the understanding that it is meant to be enjoyed family-style, like Chinese food. I pledge to do so at the expense of my desire to keep these ideas and their offspring all to myself. Thus I make my VOW OF VIRTUE.

(your signature, please)

Cut here, enclose in an envelope with a return address, and mail to ReadyMade, 2706 8th Street, Berkeley, CA 94710. All signed manifestos will be rewarded with a surprise gift.

This is the project title. It's meant to be descriptive, so don't expect anything clever.

Each chapter focuses on one raw material (metal, plastic, fabric) found in everyday objects that are ordinarily tossed out or recycled.

Feeling confused, despondent, or lost? Check this corner for chapter numbers and icons to regain your bearings.

NO-SEW MESSENGER BAG

RAW MATERIAL

c/02

NO-SEW MESSENGER BAG

$10

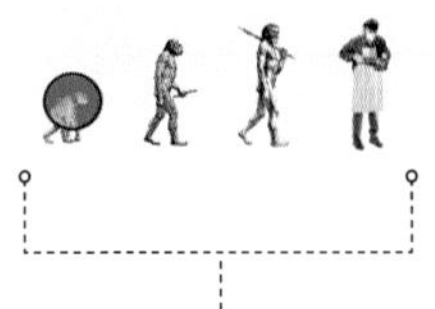

INGREDIENTS

+ Wax paper
+ One month's subscription worth of bags (25 to 30 in all—we used the large, blue *New York Times* Sunday edition bags)
+ 1 yard of black webbing
+ 2" side-release buckle
+ 2" Ladderloc adjustment buckle
+ 2 pieces of 2" by 2" Velcro

TOOLS

+ Iron
+ Ironing board
+ Tape measure
+ Scissors
+ Sharpie

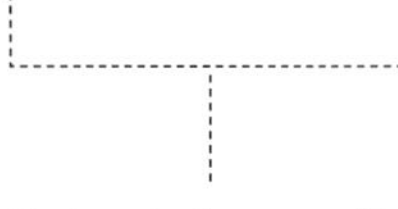

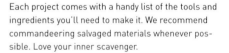

Completion times range from one hour to a couple of days, depending on how many minutes have elapsed on the stopwatch. At the 5-minute mark, you'll spend less than an hour; at the 55-minute mark, you're calling in sick on Monday.

Also, if you do not fall into one of the following categories, please reconsider doing projects yourself: Monkey (has opposable thumbs), Cro-Magnon (has tools and fire but may be clumsy with both), Drudge (has tools and basic know-how), Craftsman (knows that a "stud finder" is not a matchmaking service).

Each project comes with a handy list of the tools and ingredients you'll need to make it. We recommend commandeering salvaged materials whenever possible. Love your inner scavenger.

SAFETY FIRST

All do-it-yourself activities involve risk. Skills, materials, tools, and site conditions vary widely. Although the editors have made every effort to ensure accuracy, the reader remains responsible for the selection and use of tools and methods. Obey local codes and laws, follow manufacturers' operating instructions, and observe safety precautions. Be careful out there.

Chapter: 01

PAPER

EXTRA PULP

Paper! It's absorbent. It's flexible. It's blade-sharp at the edge and soft through the middle. It may be the most underrated invention the world has seen, delivering the message of every love letter, every declaration of war, and every fortune cookie. Paper performs the humblest tasks: chopstick holder, fire starter, wall coverer, coffee strainer. And that's just what it does at its flattest. Piled high, folded, stuck together, dunked in various liquids, or shredded, paper is capable of awesome structural feats. Some 1.5 million tons of construction products are made from paper each year, including insulation, wallboard, flooring, and sound-absorbing materials. Unsurprisingly, there are small mountains of the stuff being crumpled up every day, and it's all much better suited to a new life in a well-adjusted home like yours than going straight to the dump. Time to kick yourself to the curb to do something with all this free material. Phone books, shoe boxes, junk mail, poster tubes—they are yours for the taking, friend!

Why not just let the stuff burn up in a glorious pyre? You know why! It's made from trees. And trees are made by the magic forces of nature. The less we recycle, the more trees we have to cut down. No one wants to live in a world without forests and their columns of soft, dust-moted light. (Can you hear the bird-heavy branches swaying? The sudden bursts of choral music?) Where would all the hobbits go? Although deforestation (of the rainforest variety) is largely carried out for the purpose of creating more farmland and accruing more fuel, logging of old-growth forests (populated with trees that are two hundred years old or more) accounts for some 10 percent of virgin paper production. Americans, on average, use more than 730 pounds of paper per person per year. That amounts to nine trees, each the girth of a telephone pole and the height of a four-story building.

The good news is that trees are renewable, and American farms, planting millions of seeds each year, now contribute nearly 90 percent of the raw material used to make paper. Nearly all manufacturers include some recycled fibers in their slurry, too. But the recycling and manufacturing processes are both energy-hogging and pollution-causing, and paper cannot be recycled indefinitely—it degrades with each use, so after five or six visits to the curb it's ready to retire. As our appetite for the copier-grade sheet continues to increase (so much for the paperless office), we must remind ourselves not to take paper for granted. Use it wisely, and use it more than once.

The average American uses more than 730 pounds of paper per year.

Was that preachy? We don't mean to sound preachy. We are very much in favor of the recreational use of the white stuff. The two authors of this book alone consumed some fifteen hundred pounds of paper in the past year without any noticeable side effects. But we're always looking for new ways to make the pulp last. Here are a few projects we've dog-eared for you. Think of these creations as your own little indoor tree farm. Water your ideas. Place them in a sunny spot. Watch them grow.

A BRIEF HISTORY OF PAPER

START HERE

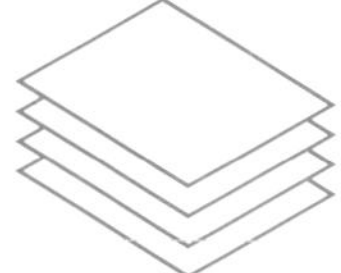

4000–2200 BC
Earliest Egyptian documents on papyrus.

1400–1300
Bones used as a writing material in China. Aristotle and other Greeks work on cowhide.

610
Papermaking knowledge is spread by Arabs to North Africa (camels do most of the work).

875
Chinese invent toilet paper and lo mein.

1100
Arab and Chinese paper is exported to Italy and Spain. European paper mill is set up by Arabs in Xativa, near Valencia (pronounced with a "th").

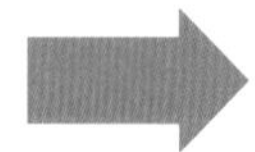

1690
William Rittenhouse and William Bradford found the first American paper mill at Wissahickon Creek, near Philadelphia. They celebrate with a cheese steak.

1700s
Rene de Réaumur stumbles upon the secret of papermaking while studying a species of wasp now called the paper wasp. These little stingers were munching on wood—not eating it, but chewing it up, then spitting the pulp out to form soft, papery nests. Réaumur didn't take the next step, chewing on wood himself, but he had uncovered the papermaking process of the future—that wood could be broken down, mixed with water and energy, and churned into paper.

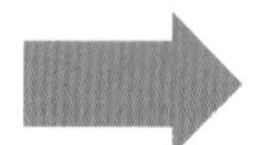

1870s
Paper is used to make a dome for the New York observatory. Meanwhile, Margaret Knight, an employee in a paper-bag factory, invents a new machine part to make square bottoms for grocery bags, which had been more like envelopes. She founds the Eastern Paper Bag Company in 1870. Go Maggie!

1877
A fifty-foot chimney made of paper is erected in Breslau, Prussia. Put that in your pipe.

1911
Thirty-two years after German chemist C. F. Dahl develops the Kraft pulping process the first kraft mill in the United States opens in Pensacola, Florida. The process pulps pine trees, the most common forest species in the United States.

1942
Paper blankets are introduced in New York City. Cold comfort.

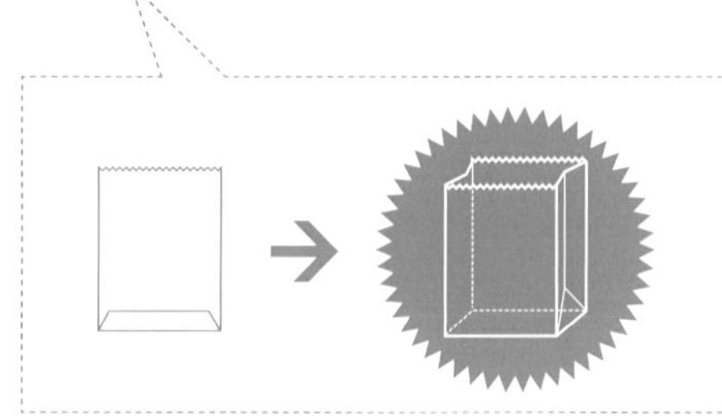

How did we get from the first cave scratchings

to the Victoria's Secret catalog?

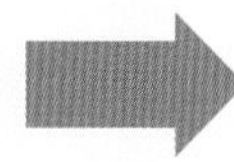

It went something like this:

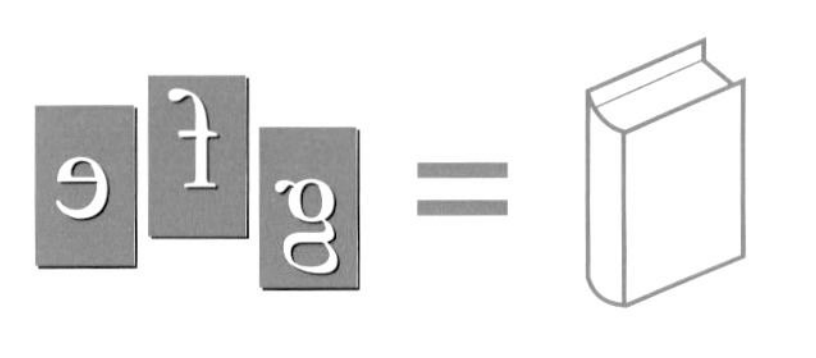

AD 105
There's only so long you can ruin your hands chiseling tablets and scrawling on rawhide before you look for a more receptive surface. So, Ts'ai Lun, a Chinese court official, invented paper. Ts'ai shredded mulberry bark, hemp, and rags, mixed them with water, beat the lot into a pulp, daubed out the liquid, and hung the pressed sheet out to dry. Then he invented chow mein.

1300–1400s
In some places, plague is thought to be brought in on rags imported by (filthy, cloddish) papermakers. Ironically, when the plague kills millions of people in Europe, plenty of cloth rags become available. Then Gutenberg invents movable type in the 1450s, making mass printing of books possible. By 1500 people start reading things besides the good book. Bring on the Renaissance!

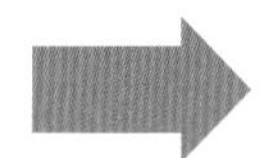

1550
Wallpaper is introduced to Europe from China. Courts cover up the bloodstains of slain nobles.

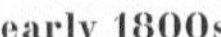

early 1800s
Nicholas-Louis Robert of France invents the Fourdrinier, a clever machine that produces paper on an endless wire screen. Fifty years later, papermakers begin successfully using wood fiber to make pulp—the process is introduced in the United States in the early 1900s.

1820s
A Brit named John Dickinson invents the first machine for pasting sheets together to cord cardboard.

1842
First Christmas card printed in England, ushering in an era of holiday-card-industry tyranny.

1863
Germans use paper to make coffee cups.

1900–1910
Appearance of the first single-use disposable paper plates and milk bottles.

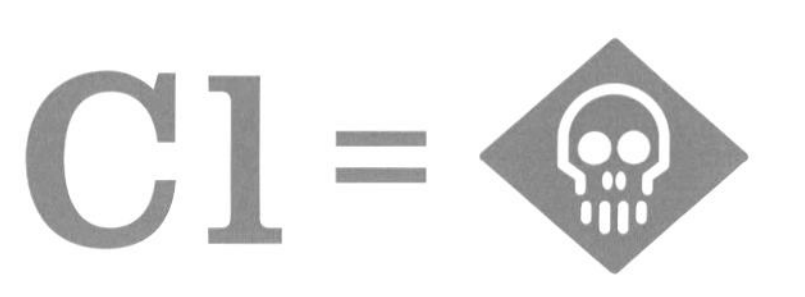

1907
Scott Paper introduces the first paper towels for use in Philadelphia classrooms to help prevent the spread of the common cold.

1960s
Save the planet, baby! Greater stress on ecology at the plants (closed loops of waste and energy production) and increased automation. First use of de-inked recovered paper, processed chemicals, and dyes, forest species in the United States.

1980s
It surfaces that elemental chlorine, the chemical used to separate and whiten wood fiber, combines with lignins to produce dioxin, one of the most potent carcinogens. (Paper mills are the second greatest source of dioxin and the largest source of dioxin in water.) Paper mills ≠ clean.

1987
First Victoria's Secret catalog published. Like paper mills, the catalog's content also ≠ clean.

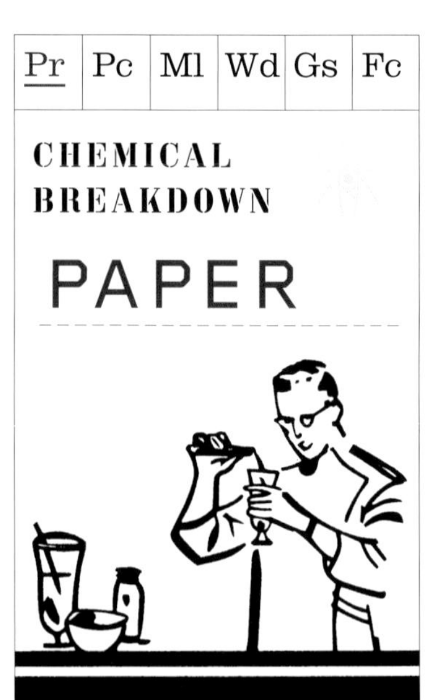

CHEMICAL BREAKDOWN

PAPER

STEP 01

TREE MUGGING

Millions of softwood pine trees are grown every year for paper harvesting. They're cut down, de-branched, and fed through a tree Cuisinart.

GOO GONE

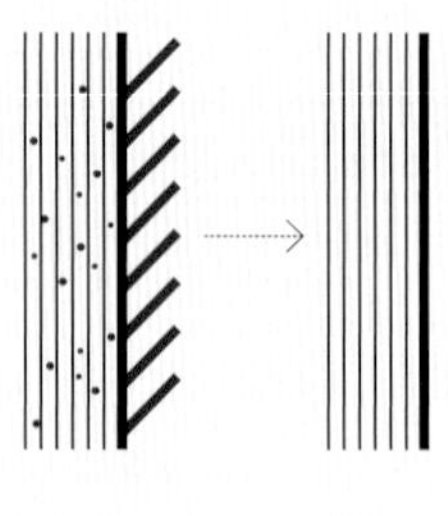

Wood fibers are chipped and stripped of their natural unguents—lignins, gums, and waxes. The chips undergo one of two pulping processes.

CHEMICAL PULPING

Developed during the latter half of the nineteenth century, this process releases the cellulose fibers in wood by dissolving and extracting the lignin with hardcore chemicals. Begs the question: How are these caustics recovered or disposed of?

MECHANICAL PULPING

Wood fibers are shredded by friction against grindstones. All lignin remains in the pulp, so the resulting paper is fairly fast to degrade and yellow. Uses: newsprint, tissue, and manila.

STEP 03

MIGHTY WHITEY

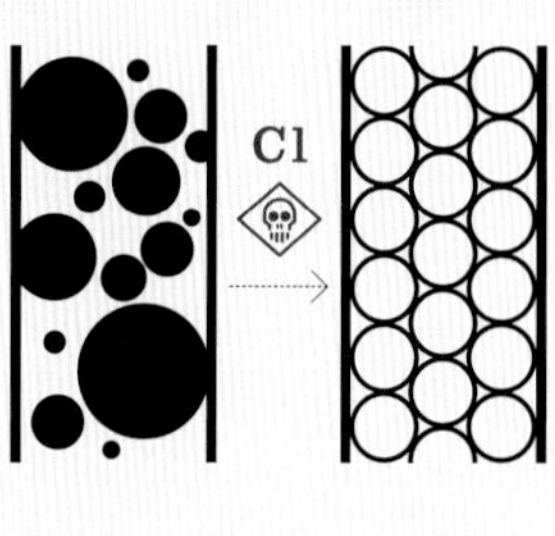

This is the nasty part. Bleaching methods include chlorine dioxide bleaching and hydrogen peroxide bleaching. Bleaching not only whitens the pulp, it also helps to increase paper strength by removing lignin left over from the pulping process. The problem: If pulp is exposed to chlorine or chlorine dioxide, the wastewater is too corrosive to go through the recovery system and ends up being funneled to a treatment system instead. Ultimate destination post-treatment: lakes and rivers.

STEP 04

A GOOD BEATING

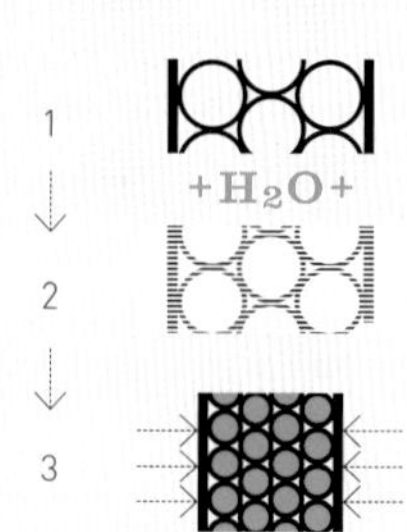

The pulp, with most of the lignin removed, is dispersed in water, then passed through beating devices—sharp cutters that slide over one another at a very high pressure and velocity, crushing the fibers and resulting in a pulp that bonds well.

FORMING/DRYING CONVERTING

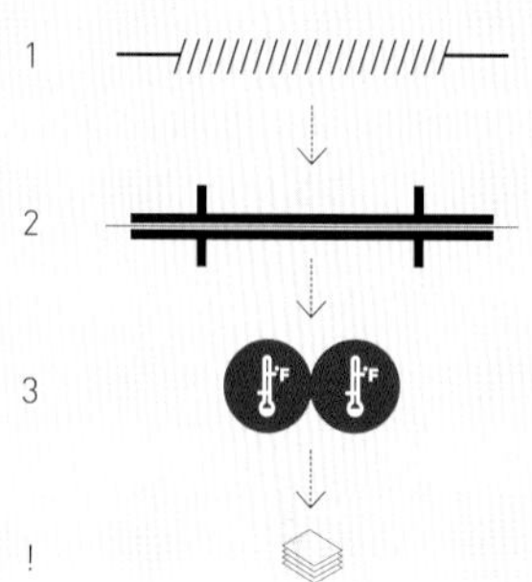

The final sheet of paper is a network of fibers that bond to one another. The pulp, thoroughly diluted with water, passes through a screen moving at a constant speed. While the water passes through the screen, the fibers are left on the screen. The wet fibrous layer then passes through presses to extract moisture, and then to drying cylinders.

THE ONE-WOMAN SHOW
MOFFETT
SAMUEL FRENCH

HARDCOVER FRAME

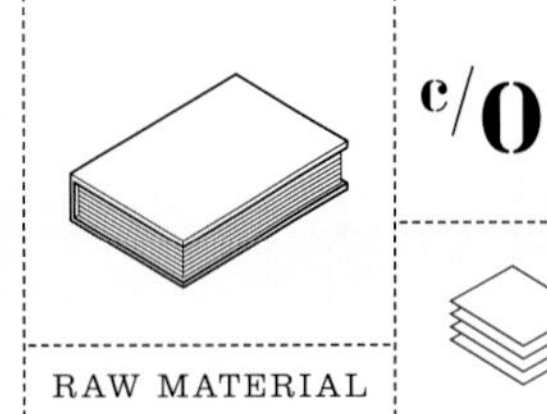

c/01

Judge books by their covers—especially if the books are lame and the covers are funny. Who can resist remainder-bin classics like *Knitting with Dog Hair*, *G.I. Nun*, and *It's a Small World After All: Four Billion Years of Microbial Evolution*? Such tomes may no longer be bedside material, but their goofy titles and cloth-bound jackets are too good to be glibly tossed away. Preserve these discards by creating a double-sided picture frame. While most photo displays are mere scaffolding with little to recommend them, old-book frames let you tell the truer tale, the bigger picture.

Percentage of office paper that ends up in landfill: **40**

I was certain the man had robbed me, after thrusting his hand into my pocket and scuttling off the train. But when I felt for it, my wallet was there, along with a fistful of tiny bits of paper. When put together, the note divulged an address, and above it, the simple phrase: "Seeing you, I came apart."

HARDCOVER FRAME

$10

INGREDIENTS

+ Old book with clever title
+ Cardboard
+ Book-sized sheet of Mylar
+ Photo corners
+ Book-sized pieces of decorative paper

TOOLS

+ Utility knife
+ Pencil or marker
+ Ruler
+ Glue

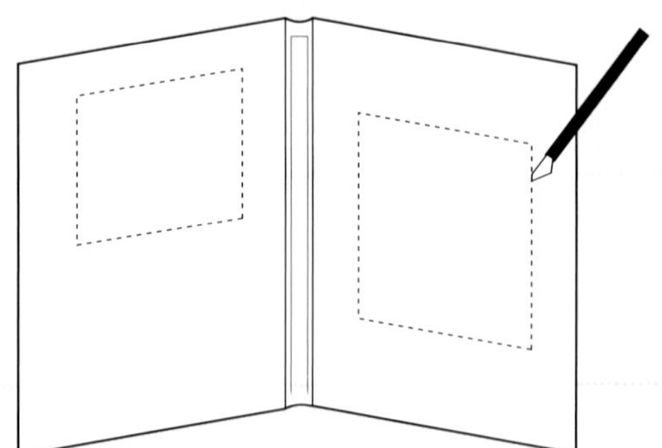

FIGURE A

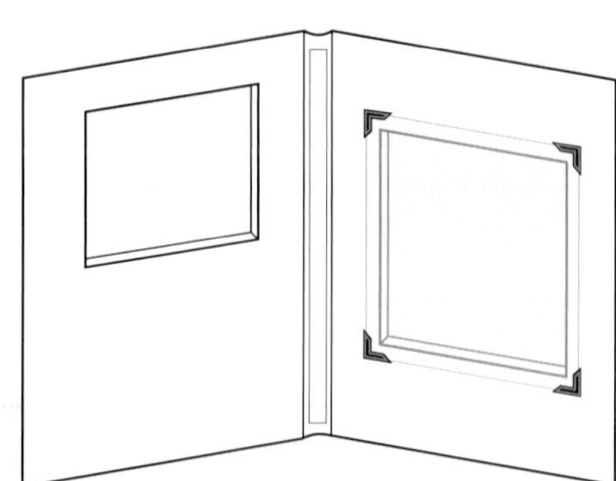

FIGURE B

1 Using the utility knife, carefully remove the "guts" of the book at the spine in one large piece. Keep the cover in one piece.

2 Cut a piece of cardboard to add structure to the now-limp spine.

3 Maintaining the integrity of the title on the front cover, use a ruler to outline a rectangular frame opening on both the inside of the front cover and the inside of the back cover.

4 Use the utility knife to cut where you've marked (Figure A).

5 Brush a light coating of glue over the raw edges on the front and back covers around the frame opening.

6 While the glue is drying, cut the Mylar into two pieces—each $^1/_2$" larger on all four sides than the frame openings.

7 Place the photo corners inside the covers to hold the Mylar in place in front of the frame openings. The Mylar will protect your photos (Figure B).

Global industry responsible for the fifth-largest consumption of energy: Paper

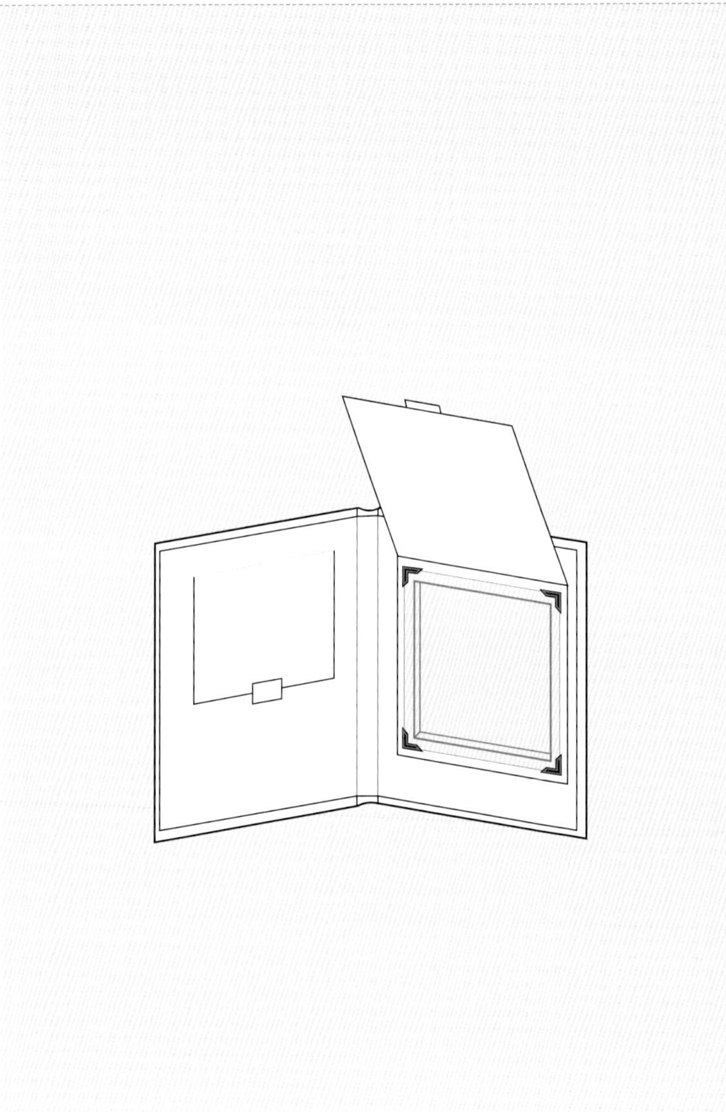

FIGURE C

8 Glue the outer edges of the decorative paper to the backside of the cover without adhering it to the Mylar.

9 Cut three sides of the decorative paper perfectly to size around the Mylar, leaving the top edge uncut. Adhere another small piece of paper to allow the flap to self-tab (Figure C).

10 Trim the photo for display to the size of the Mylar.

11 Open the flap and place the photo in the photo corners to secure in place.

Biggest consumer of paper per person in pounds in 2001: **Belgium**

FIBERSPACE

Learn about paper, then buy some!

Paperworks *(www.paperworks.biz) Download paper project instructions.*

Vision Paper *(www visionpaper.com /kenaf2.html)* Where to get your alternative Kenaf-based paper.

Rethinking paper *(www.rethink paper.org)* An ecologically-minded nonprofit that charts a course of action for reducing your wood use.

Mr. French paper *(www.mrfrench. com)* The best-looking paper retailer we know—the catalog alone is a work of art.

Paper.com *(www.paper.com)* The superstore.

Sam Flax *(www.samflaxny.com/ paperstationery.html)* Flax has a bottomless collection of tools and materials for your paper projects.

Kate's Paperie *(www.katespaperie. com)* Seller of fine boutique papers, journals, and scrapbooks.

Printed Matter *(www.printed matter.org)* New York–based nonprofit dedicated to the promotion of handmade books and other arty paper products.

Book Arts Web *(www.philobiblon. com)* Clearinghouse of handy information and links for book artists and book art appreciators.

Express
Express
Express

FEDEX CD RACK

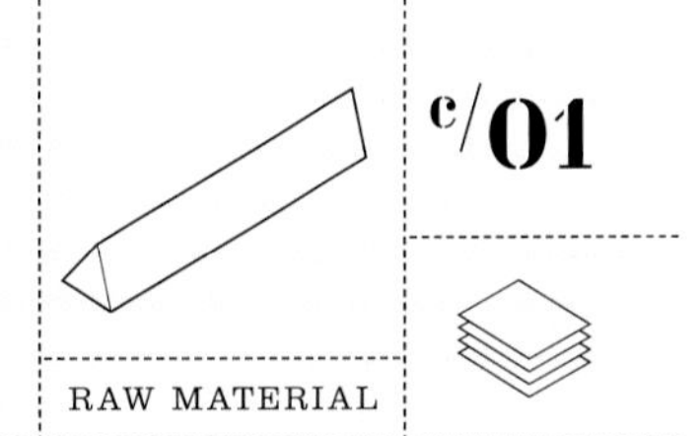

Compact discs: the great storage problem of our day. The used sections of music stores are bulging to capacity. Indian reservations won't take them. If dumped at sea, they'll just bob upon the waves. Plus, discs might come back into fashion. (We can hear the kids of tomorrow saying, "You have the original CD?!") If we've convinced you at all to keep your collection, you'll need an expandable storage solution. Most racks have that style-starved "I'm not a sculpture but I play one on TV" look. Here's an overnight idea—cut slits into FedEx tubes and hang as many of them as you need to house your listening library.

When I was very young, I once wrote a letter to Santa Claus and insisted that I send it myself, stamp and all. I had the unfortunate luck of fetching the mail the day the letter was returned. I assumed that this was standard treatment for Jews who wrote to Christian icons.

Number of pounds one year's worth of the *New York Times* weighs: 520

FEDEX CD RACK

$5

INGREDIENTS

+ Triangular FedEx poster tube
+ Picture-hanging claw

TOOLS

+ Ruler
+ Pencil
+ Heavy utility knife

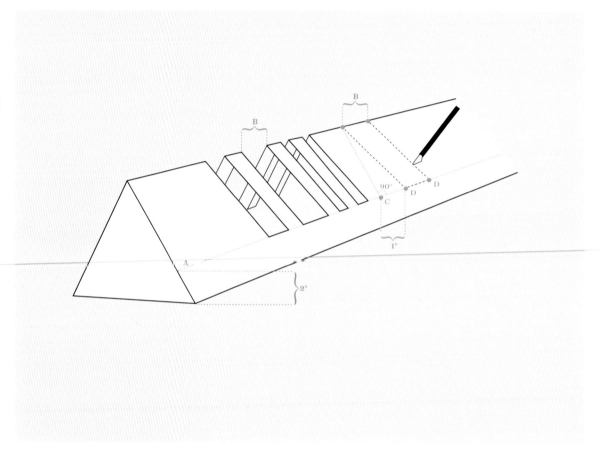

1 Lay the double-walled cardboard side of the shipping tube down on the table. This will provide the sturdiest backing for the rack.

2 Measure 2" from the bottom on both sides and dot with light pencil marks down the length of the tube. Pencil a line to connect the dots (A).

3 Determine how many CDs you would like your rack to hold, and in what configuration (i.e., in singles or groups of two, three, or four).

4 Measure and make pencil marks at your determined configuration on the apex of the triangular length of the tube (B). Make sure to leave a couple of inches between the first opening from the top and bottom for stability. The following guideline measurements provide snug openings for jewel cases:

1 CD: 13/32"
2 CDs: 13/16"
3 CDs: 1 7/32"
4 CDs: 1 20/32"

Name of the first paper merchant in America who helped found eighteen paper mills in Virginia and surrounding areas: Benjamin Franklin

REDUCE YOUR TREE DEPENDENCE

Amazing Savings!

Businesses that use vats of paper products are key in the effort to preserve forestlands. The United Parcel Service, for example, has potentially halved the packaging used for overnight delivery by introducing boxes that can be used twice. You captains of industry who are reading this: Follow UPS's example!

A FEW TIPS: Use and reuse rags vs. paper towels; save one-sided paper and make scratch pads; use cloth bags to carry groceries; use old magazines and newspapers to wrap gifts; bring reusable containers for takeout; buy bulk foods to reduce packaging consumption; if you own a Xerox machine, change your default setting to double-sided; use e-mail instead of paper memos; post or circulate materials rather than printing copies for everyone.

DUCHAMP'S CORNER

20 Alternate Uses for Your Old Newspaper

Packing filler
Confetti
Bug swatter
Last minute rain hat
Toy boat
Pet potty trainer
Paper flowers
Window wiper
Makeshift sponge
Megaphone
Face hider
Fan
Gift wrap
Floor tarp
Paper airplane
Fire kindling
Window shade
Doggy fetch toy
Spit balls
If you must, makeshift toilet paper

5 Lightly pencil a straight line from your marks at the apex to the line along the length of both walls of the tube (A)‡(B). This intersection will be point (C).

6 Now measure 1" behind your (C) marks on both walls and mark (D). This is will be where you will begin the cut that will form the foot of your openings. This step is very important since it provides the slant of your CD openings.

7 Cut with your utility knife from mark (D) to mark (B).

8 Now cut between the (D) marks to create your openings.

9 Insert and secure two claw picture hangers close to the top and bottom of the back of the rack.

10 Hang on the wall and load up your collection.

The New York Times
28 SMALL SPACE SOLUTIONS
MAKE YOUR PLACE BIGGER
TINY KITCHEN, HUGE FEAST

POSTER-TUBE MAGAZINE TREE

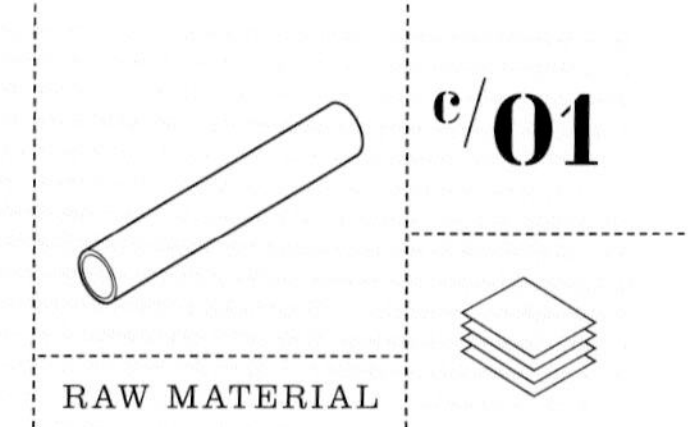

Your plants keep withering. You water them. You think loving, mineral-enriched thoughts about them. You move them into shadier spots. Nothing works. Part of you doesn't want to learn how to make things grow; there's an appealing riddle to the great outdoors—how sun and rain join ranks to conjure saplings out of soil—and you're fine not knowing all the answers. So, in place of the high-maintenance organic stuff, try a poster-tube tree. It has *Through the Looking Glass* appeal, like a mushroom with a technicolor cap. Save your mailers, or get new ones at the post office, and plant some periodicals.

Pounds of paper the average United States citizen consumes in a year: 749

Pounds of paper consumed by people in developing countries: 40

POSTER TUBE MAGAZINE TREE

INGREDIENTS

- 6 poster shipping tubes of various diameters (6", 5", 4" recommended)
- 1 1/2" nails
- 1/8" rivets
- Trash-can rubber bands or twine

TOOLS

- Pencil
- Hacksaw
- Ruler
- Hammer
- Rivet gun

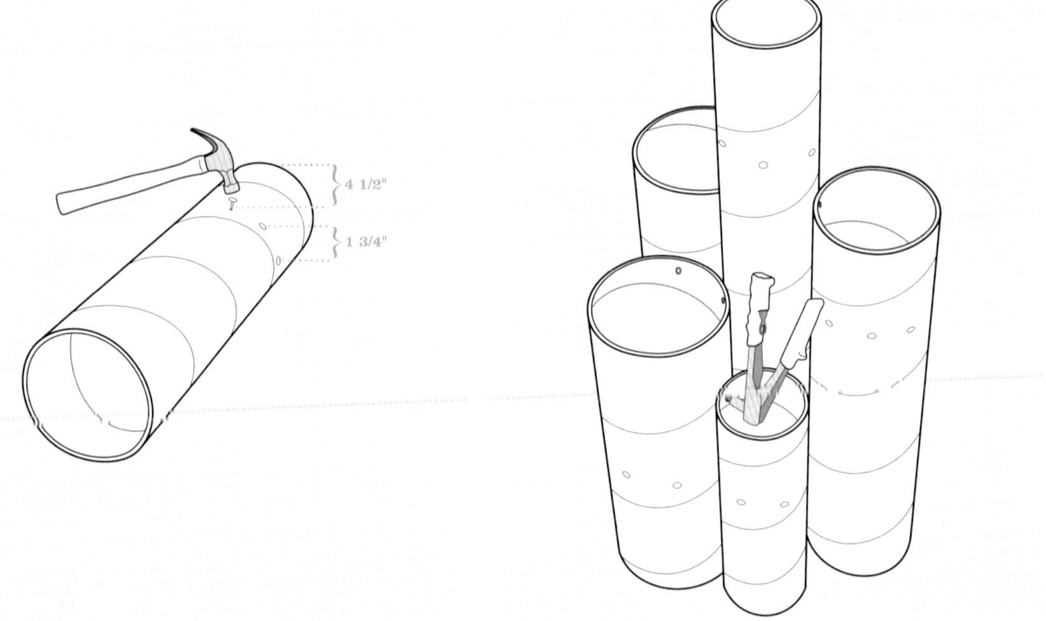

FIGURE A

FIGURE B

1 You will want to vary the heights of the tubes for best display. Measure and mark your cuts with a pencil. Think of a sprouting tree. Our tallest tube is 30" high and our shortest is 13" tall. Note that the widest diameter is good for newspapers and your fatter fall issues.

2 Cut the tubes straight across with the hacksaw.

3 Measure 4 1/2" down from the top of your tubes and lightly hammer in a row of nails around the circumference of the tube, 1 3/4" apart. This will keep your magazines from falling to the floor of the tube (Figure A).

4 Group your tubes in a handsome arrangement and temporarily hold them in place with a large rubber band or string.

5 Rivet the tubes together 1" from both ends to fix arrangement (Figure B). Remove the rubber band or string.

6 Roll up your magazines and newspapers and plant them in the tubes. Place the arrangement in a corner suited to reading.

Percentage of all harvested trees cut down each year to make paper: 35

He took books into the bath with him—steeping in their stories for hours as his toes pruned. When, on occasion, they fell into the steaming water, he'd tear out the sodden pages and hang them, one by one, from the low laundry line in the courtyard. This is how I read my first books, the pages dripping, swooning in the wind.

PHONE-BOOK FURNITURE

The bulging directories that get plunked down on doorsteps each year are outdated before you've made it halfway through C: Chinese Takeout. We were bent on finding some new, better use for them than padding your recycling bin. But our idea of a simple paper-based table got a little out of hand. You'll find a better way to use yours; just break it to us gently when you do.

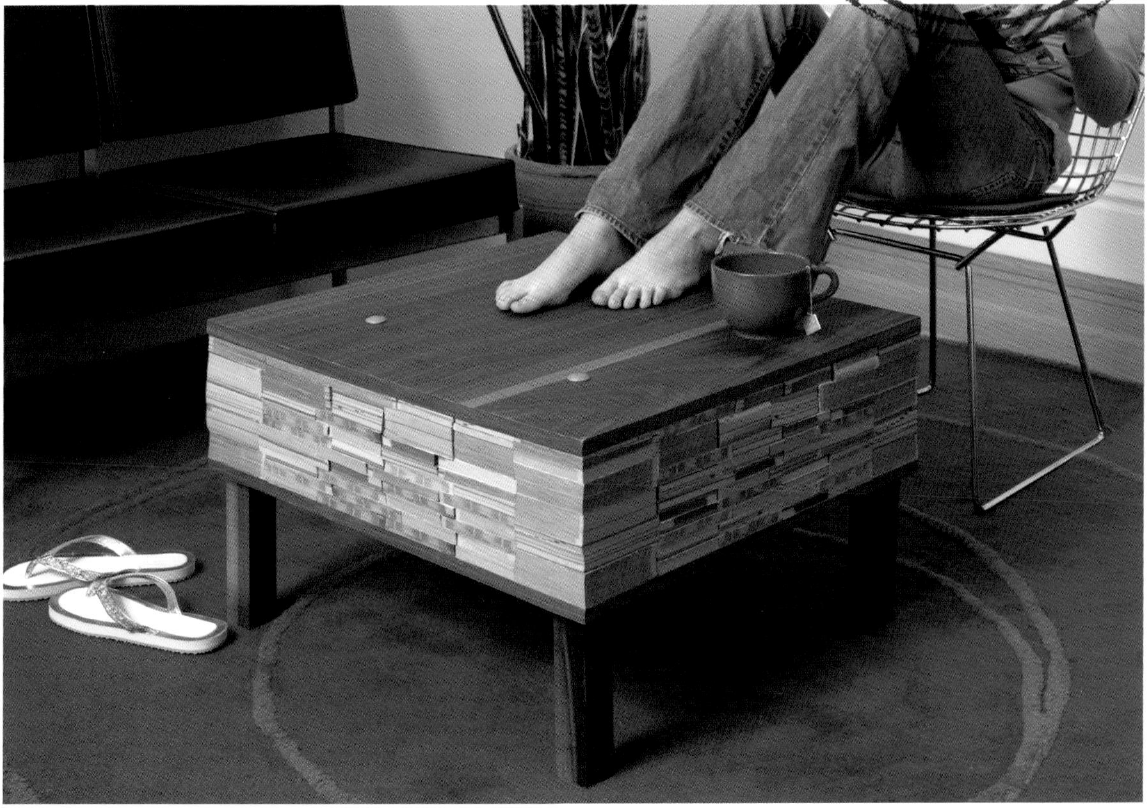

YES

1. Phone-book bricks sounded like a good idea.
2. They have a pleasing, Missoni-esque striped edge.
3. The compression of long bolts holds the whole thing together.
4. The bolt heads look like cushion buttons.
5. Keeps fourteen Yellow Pages out of circulation.

NO

1. Gluing the cut edge for durability and stiffness is a tedious process.
2. You can't drill through a phone-book brick without twisting the innards to hell.
3. The reuse material looks like it took a backseat in this project.
4. There was no way to achieve consistency in the phone-book block size.
5. The gaps don't exactly inspire confidence.

DON'T BE A HALLMARK SAP. SHOW YOUR AFFECTION IN A WAY THEY'LL NEVER FORGET.

How to Pen the Perfect Love Note

The most important thing you can do with paper is write on it, especially if your intent is to express a pretty sentiment or deep desire. In short, a love note. As we have now come of age, traveled widely, drank much, and (in this order) been boldly pursuant and pursued through the years, we've collected an almanac of ways to achieve that end of ends, namely, the getting of some. Here are a few ways to—as the Beastie Boys would have it—get crafty:

1. All notes should be made in longhand, preferably with arcane and decadent tools such as a feathered quill, drips of scented oil, pressed flowers from your English garden, and poured-wax seals. If that is simply not possible, for God's sake learn how to write a proper electronic missive. When it comes to e-mail, expressions of love should be either very short or very long. If you decide upon the very short variety, try a subject line with an elliptical phrase that continues in the body of the e-mail, such as:

   Subject: "You make a blind man . . ."
   Body: "see."

   OR

   Subject: "She's a candle . . ."
   Body: "to your sunshine."

2. If you're going long-form, it's best to start out in a humorous vein, then proceed to matters of consequence. Make an inside joke about last night's fumbling first advances or about spending the day in some holy place, praying that your sinful thoughts might be expunged (which would take a good flogging, no doubt). Then move on to the good stuff—how you have stopped eating, how you can't think about your beloved without your chest pounding, how you've already wandered down a thousand roads with him or her in your mind.

3. Metaphors and idioms are of some use: "Your smile is wider than Texas, warmer than rum on a winter's night." "You're like Christmas on a stick." "You light me up like a switchboard." "Why would I want sand when I have the beach?" And so forth. Always compare him or her to great extremes. Hyperbole is the rule.

4. Make a mix tape. Record your own voice between songs describing certain conditions for listening (while eating a mango, jumping up and down, leaning out the window at night).

5. Create a treasure hunt. A few ideas for where to leave clues: at the library, tucked inside the pages of a book; in the park carved into a tree trunk; under a pillow; at a bakery where you've bribed someone behind the counter to bake something with a message in it.

6. Write a song. No matter how poorly you sing, leave it on your intended's answering machine. Or, alternatively, leave several refrains of an Al Green ballad—no explanation necessary.

7. If you develop a crush on someone in a restaurant, leave a brief message or flattering phrase (see No. 2 above) on a napkin and have your waitperson deliver it with a drink you spring for. If your encounter occurs on a bus, write your phone number on a transfer and hand it off as you disembark.

8. Use flowery, semi-opaque language. There's a reason Shakespeare's ladies fall under the spells of despots and swoon for asses.

9. Write in sign language. Get a book of hand signs for the deaf, cut out the words and letters you need to write your note, and make them do the work of rendering your feelings audible (see below).

10. Make a T-shirt. Get iron-on letters and spell out I LOVE YOU or MARRY ME, then wear the garment on a date. Or buy a button-making machine. Make buttons that make clear your intentions and wear them on a blank white shirt.

11. Break something. Use rub-on letters to inscribe the reasons you cherish your intended onto a piece of marble or other friable surface. Smash it into pieces and place it in a gift box. Give him or her the archaeological pleasure of putting it back together to decipher your message.

12. Wear a RESERVED FOR . . . sign. Get the kind that are placed on seats at concerts and write in your beloved's name. Wear it on the back or the front of a T-shirt at a public event.

13. Tear a photograph of yourself in half and put the two pieces in two lockets, one for you and one for your beloved. Your halves are whole only when united.

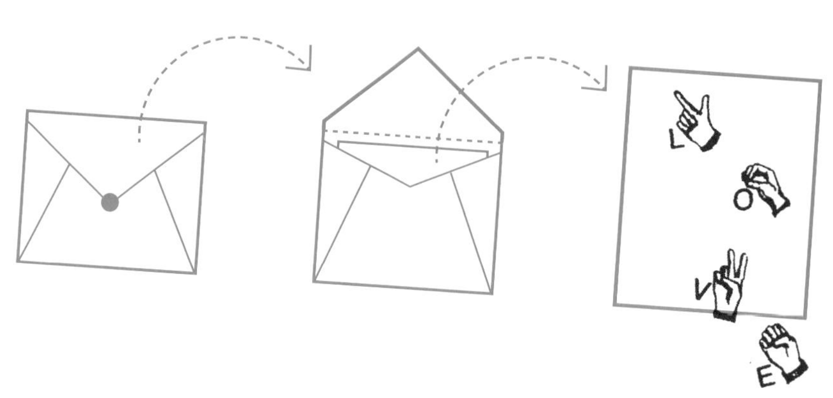

redibly unique
USDA Prime ar
market

SHOPPING-BAG RUG

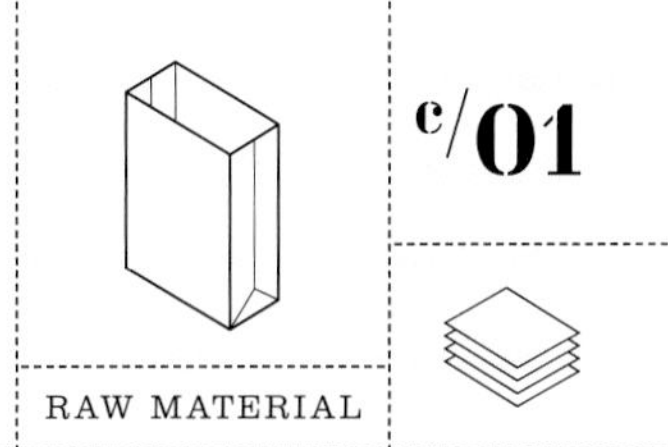

You try to be good. You know from Hillary that it takes a village to accomplish real change. To that end, you gathered a small fleet of canvas totes to haul home groceries. But when you're out strolling and the Buy It! impulse hits, you can't deny the satisfaction of carrying home a heavily lacquered shopping bag. The thick paper is too hardy to throw out, and perfectly suited to making a woven rug. There are similar leather versions out there—trés elegant, trés expensive—this version is closer to the tribal village that preserves resources through reuse. Now if only someone would go hunt down dinner.

What the first newsprint was made of in the late 1800s: **Clothing rags**

My mother called it her "worry bag." It was a thinning brown paper sack, with a drawstring at the top. Whenever there was trouble with my father or with us, she'd sit it next to her with one hand inside, working her fingers over something. We never dared ask what was in there, but it always calmed her down.

SHOPPING-BAG RUG

$45

INGREDIENTS

+ 15 large, heavy-paper shopping bags (for a 4' x 6' rug)
+ Clear packing tape
+ Carpet tape
+ Varathane (semi-gloss)
+ Rubber carpet skid

TOOLS

+ Industrial scissors
+ Ruler
+ Utility knife
+ Paintbrush

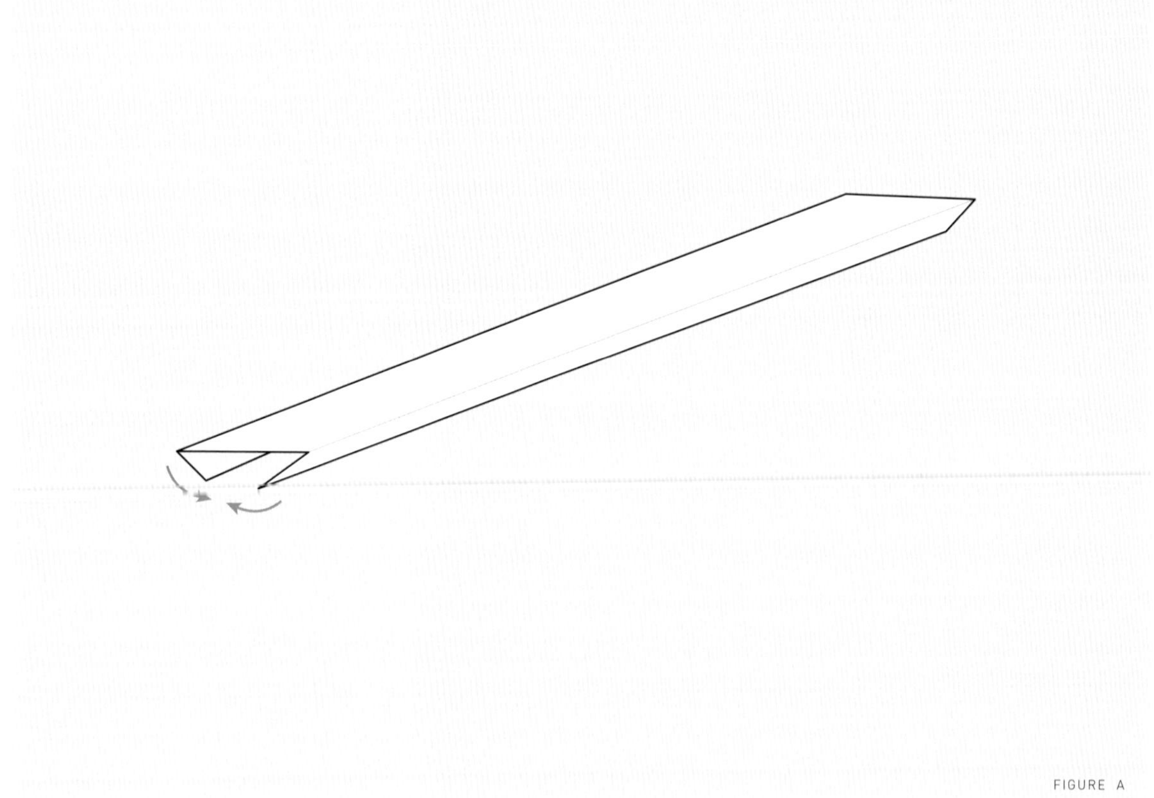

FIGURE A

1 Dismantle the shopping bags by removing the handles and undoing the seams so you end up with one large flat piece from each bag.

2 Make one to three different widths of strips (1", 2", 4") by cutting and folding the flat bags (Figure A).

3 Organize your strips by width. Strips will vary in length.

4 Clear a large workspace and connect two or more strips of the same width with clear packing tape to form one edge of your rug.

5 Connect two or more strips of the same width with clear packing tape to form an adjacent edge.

6 Tape these two edges together to form an "L." This will anchor the rug.

7 Begin to basket weave the strips, row by row, progressively from the anchor corner you just made. Make sure to alternate strip widths by entire rows, never within a row. Secure the ends of the strips to the edge of the anchor strip by folding them over and taping them down as you progress (Figure B).

Where the great leaders, past and present, of the paper industry are inducted each year: Paper Industry Hall of Fame, Appleton, Wisconsin

FIGURE B

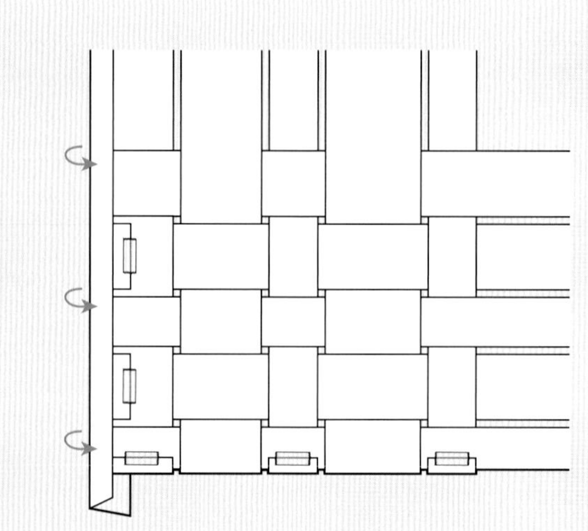

FIGURE C

8 After weaving a row, periodically tighten the weave by closing up any spaces between the rows from the anchor corner out.

9 Finish the edge by adhering carpet tape to the underside of the strips you selected for the edge and folding the strips along the edge of the rug (Figure C).

10 Once weaving is complete, apply two coats of Varathane in a well-ventilated area. Allow your rug to dry completely.

11 Adhere the meshlike rubber carpet skid to the bottom with carpet tape to keep it from slipping.

12 Put on the Village People and cut a rug.

TIP: Take note of color, pattern, and durability in your bag selection process. The boutique type, with soft rope-like handles, works best.

Percentage of energy that recycling paper uses vs. manufacturing it from virgin timber: **60**

Newspapers shape h
William Randolph He
New York Journal pic
sunk in the Havana h
the Spanish America
it was a coal fire tha
down, not "Spanish t

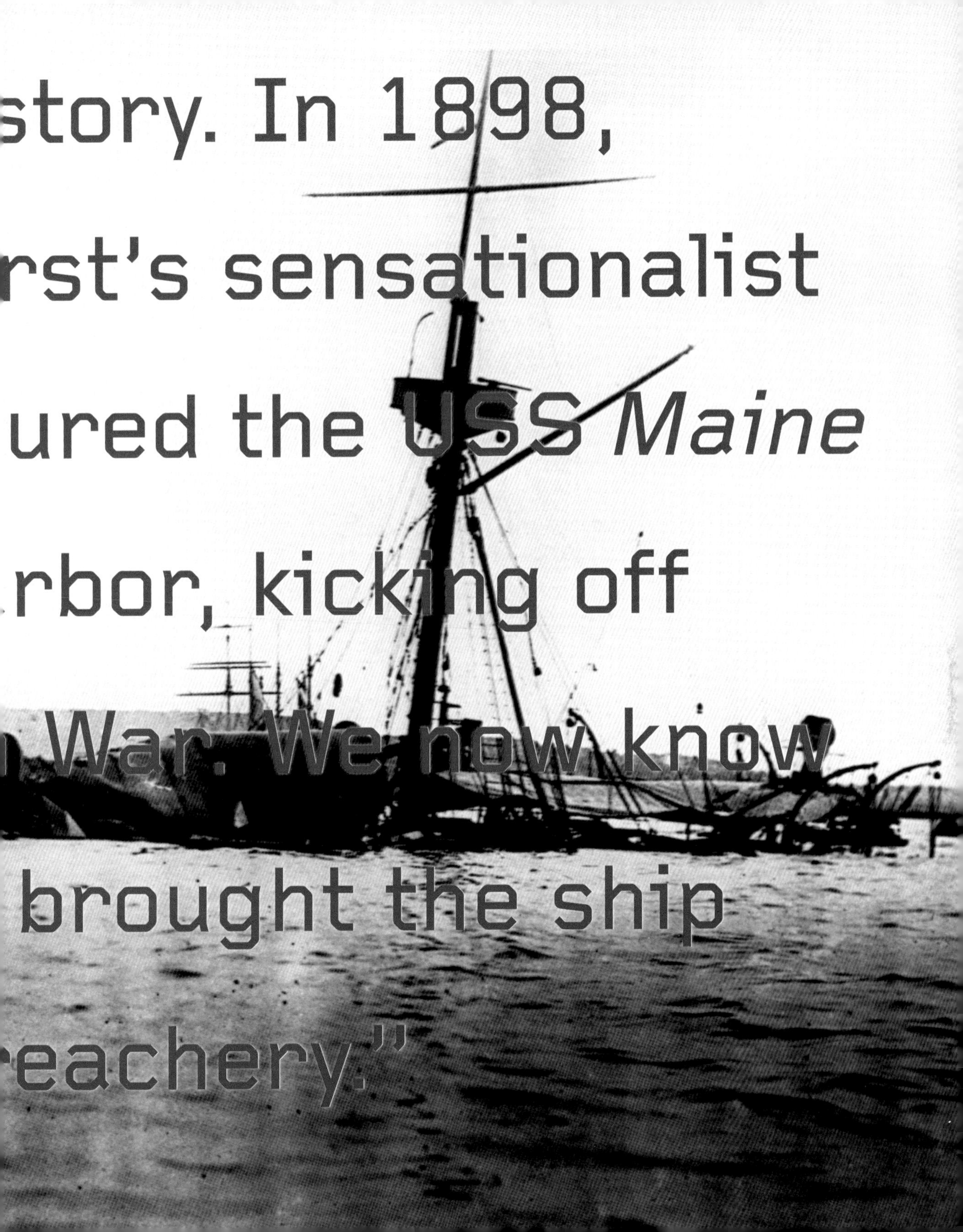

story. In 1898,

rst's sensationalist

ured the USS *Maine*

rbor, kicking off

War. We now know

brought the ship

eachery."

SHOE-BOX SHOJI SCREEN

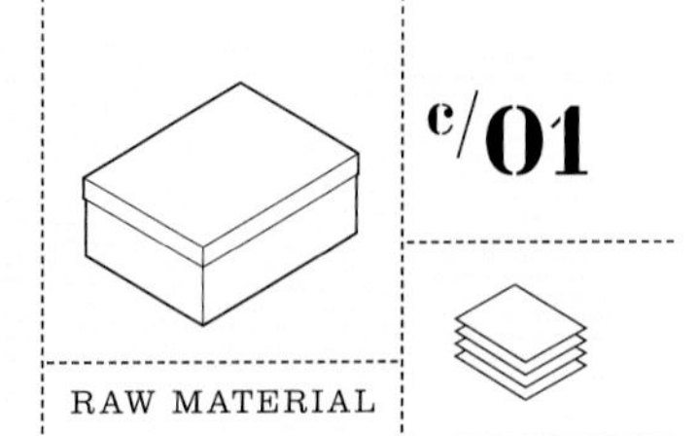

Have you seen the sneakers they're selling lately? The stitched layers of dyed leather, the biomorphic curves, the space-age sweat-wicking fabrics, the all-around pimp-daddyness? Owning a collection of kicks means two things: lots of empty boxes and less closet space. This three-in-one rolling wardrobe, modular storage unit, and room divider solves both problems. It houses your trainers inside the very container they arrived in while clearing space for all the other poor soles that are getting crowded out. Fold it closed like a suitcase, or leave it open like a screen to mark your territory.

Year the Scott Paper Company introduces toilet tissue on a roll: 1890

SHOE-BOX SHOJI SCREEN

$50

INGREDIENTS

+ 18 shoe boxes (using similar sizes yields the best result)
+ 1/8" rivets
+ 1/4" plywood
+ 3/4" 20-gauge finish nails
+ 8 mini-casters and screws
+ Hinges
+ Wood stain (optional)

TOOLS

+ Ruler
+ Utility knife
+ Rivet gun
+ Hammer
+ Hand drill
+ Paintbrush (optional)
+ Rag (optional)

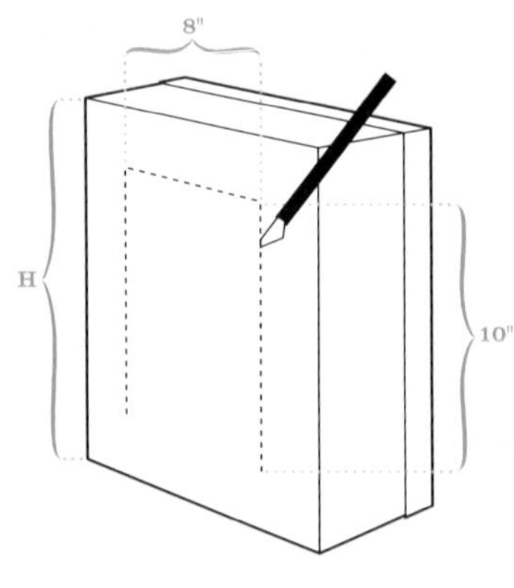

FIGURE A

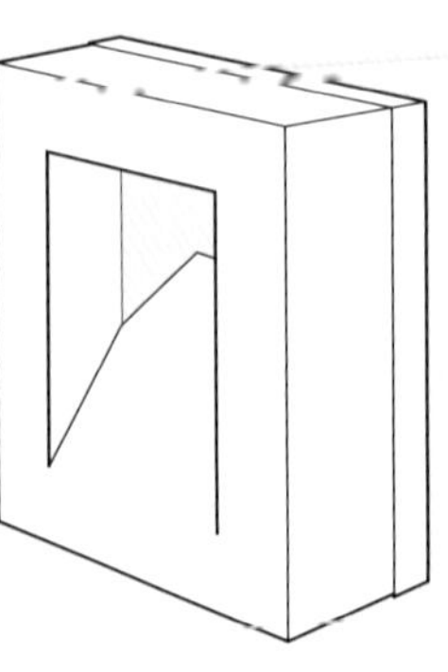

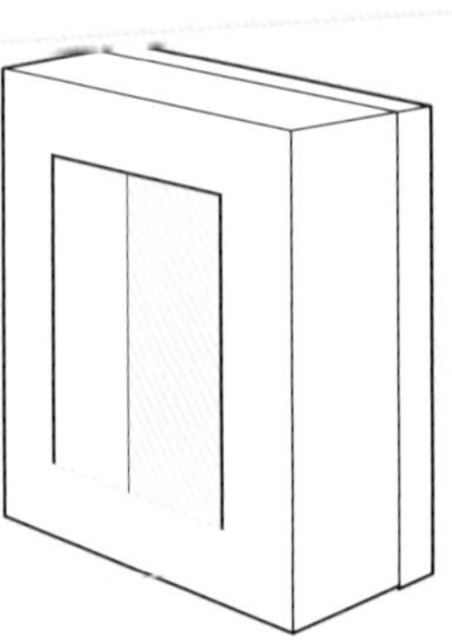

FIGURE B

1 Dust off nine of your favorite cardboard shoe boxes and sort by box height into groups of three. These will form the rows of your wall. (The walls do not need to share the same dimensions, but rows should contain boxes of similar height.)

2 Measure and cut out the right, top, and left sides of a centered opening, 8" by 10", in the bottom of each shoe box (Figure A). Remember to leave the foot of the opening uncut.

3 Lightly score the outer edge of the bottom without cutting the cardboard flap off and fold inward to create a downward self-ramp in the interior of the box (Figure B). This will keep your shoes upright and inside the shoe box.

4 Arrange the shoe boxes in the desired rows with alternating openings facing forward on the floor. This will ensure that shoes are on display on both sides of the screen.

5 Rivet the side walls of the shoe boxes together near the four corners.

6 Rivet the rows together at two points on the floor of each box in the

What people used before the advent of toilet paper: Newsprint; corn cobs; Sears, Roebuck catalogs; mussel shell; leaves; sand

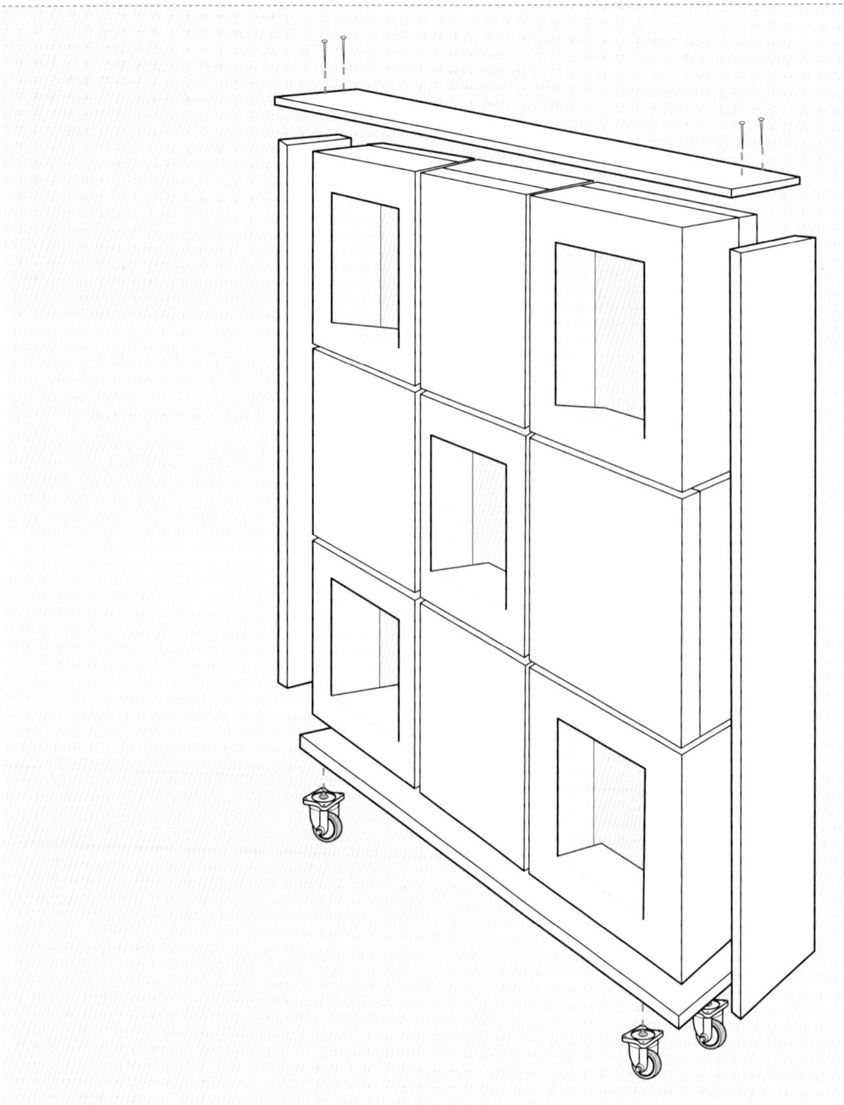

FIGURE C

top row to the ceiling of each box in the middle row. Repeat to secure the middle row to the bottom row.

7 Measure out and cut the dimensions of the ¼" plywood walls you will need to create a casing for your shoe-box wall (Figure C).

8 Use your finish carpenter nails to secure the wood walls together.

9 Screw four small casters to the bottom of the wall.

10 Repeat steps 1 through 9 to make another wall.

11 Screw two hinges to connect the two shoe-box walls together.

12 You can stain the wood for a more finished look.

13 Load up your sneakers.

TIP: You need four pieces of ¼" plywood for each wall—two of the same size for the top and bottom, and two of the same size for the sides. The dimensions depend on the shoe-box sizes.

Percentage of the world's population the United States holds: 5

Percentage of the world's paper supply Americans consume: 30

this is not a project

№01 PAPER

THE GREAT PUBLICATIONS OF OUR DAY WERE ONCE A MERE IDEA. HERE'S HOW TO GET STARTED.

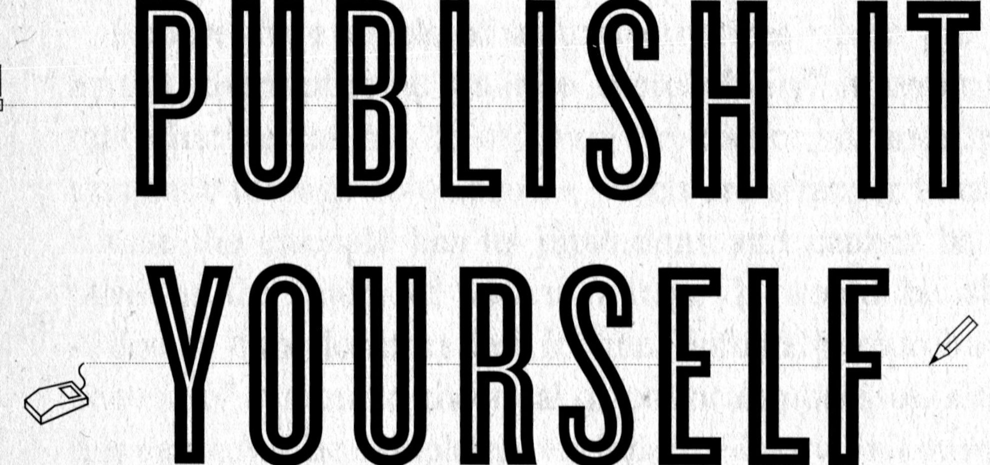

MAKE	BOOKS,	PAMPH	LETS,	AND	OTHER	ARCANA
1	2	3	4	5	6	7

STORIES WHORL around your head and make a mess of your hair. Lines of dialogue waft to the ground, leaving a trail behind you as you walk to work. You, my friend, are a good candidate for self publishing. Forget jockeying your way into the literary scene, with its shabby gentility and Truman Capote party gab. Getting into print the old-fashioned way can be a painfully long, hand-wringing process. Better to get your thoughts to market while they're still farm-fresh. Once you've taken your first stab, opportunities will open up like curtains. We're not suggesting you use one of those print-on-demand services; we're talking the raw deal—getting inky at Kinko's, or taking your job to a professional printer and hitting the pavement to distribute it yourself.

Here are a few ideas to get you started. We highly recommend using today's exhibitionist tool par excellence—the blog—as a primer. Blogs are easy to set up and cheap to maintain. They are the perfect public sounding board for you to bounce ideas off of.

BLOG STARTER PACK

Best tools for beginners: Blogger, Movable Type, WordPress, TypePad, and LiveJournal. Blogger (new.blogger.com) kicked off the Web journaling craze in 1999 and now belongs to the search baron Google. The software is easy to use: Follow the setup wizard through a four-step process, and you're ready for prime time. TypePad (*www.typepad.com*) lets you add pics from a photo-sharing site like Flickr (www.flickr.com). You know what they say, a picture's worth . . .

Usage statistics are one of the major advantages of online publishing. You can check the number of people who browse your pages each day and get a list of the sites that link to yours. Much of the blog software will do this for you automatically. Check Technorati (*www.technorati.com*) to see who's linking to you.

Spend at least a month posting regularly and see how people respond. If you're on to something, you'll know it by the leap in traffic on your site. Once you've worked the kinks out of your writing in the blogosphere, it's time to go to print. Most magazines, literary reviews, and autobiographies of note start small, with the kernel of a story. The slim, stapled chapbooks that you see on the newsstands at independent bookstores are affectionately known as zines. Many of these black-and-white affairs start out with modest intentions and grow into major cultural influences (cf. *The Paris Review, The Baffler, McSweeney's, Giant Robot*).

"The classic tools for these 'memos from the front lines' are a copy machine and staples," says Mimi Zeiger, founder and publisher of the architecture zine *loud paper*. "Voice is at the heart of zine publishing: from hastily Xeroxed rock-and-roll manifestos to occasionally heady architecture journals, it is irrepressible content—something that must be said. Voice is the justification for all DIY publishing." *Loud paper* created a savvy, urban culture voice, and launched Zeiger's writing crareer.

ZINE STARTER PACK

***Biosfear*, an urban-planning** and environmental zine created by Rob Fracisco, started out with Rob doing door-to-door distribution. Here's how he got it into circulation: "After I found out how much printing a zine really costs, I went and got a job working for a printer that ran a state-of-the-art four-color press. When the saddle-stitched booklets came out, I went to concerts, art openings, and other events to sell enough copies to cover my extremely discounted production costs. Also, I sold them at any local bookstore or comic-book shop that would take them. When traveling, I always took a stack with me no matter where I was going. In the process, I made a lot of connections that helped me spread a wider distribution web for the next issue."

If you don't see a job at a printer in your future, here's a microeconomics primer for zine publishing: Plan on spending $250 to produce a thousand black-and-white copies. If you charge $1 for each issue, you should make that back in no time. For a four-color job, plan on closer to $500, and charge $2 to $3 per issue.

Graduating from zinehood into full-fledged magazine or book publishing is less of a leap than most people think. There's now a whole Starbucks full of bloggers with book deals—Wonkette's Ana Marie Cox, Choire Sicha formerly of Gawker, the list goes on. And, in the magazine world, *Surface* began very humbly as a stapled broadside handed out at parties and clubs for free. (It's now a thick, glossy newsstand mainstay.) The first issue of *McSweeney's*, Dave Eggers's quarterly literary concern, was a sparely designed paperback he mailed in manila envelopes to friends with "Hi" scrawled across the envelope's seal. The journal's success spawned an imprint, and he now publishes an entire library of fiction and art books.

Even *ReadyMade*, the magazine that inspired this book, started as a garage publishing operation, with two wild-eyed founders working in a leaky warehouse in Berkeley. The preview issue was just twenty pages and mailed by hand to five hundred people who'd signed up for a trial subscription. Start small, but think big.

c/01  

Su Mo Tu We Th Fr Sa

Why Don't I Do This Every Day?

TESTIMONIAL # 0034

NAME: **Adam Brodsley**

FROM: **San Francisco**

OCCUPATION: **Graphic Designer, Dog Walker, Real Estate Speculator**

Adam, why don't you do these things every day?

Paper cuts, paper cuts, paper cuts! I can't seem to avoid them. I guess I'm just not that good at handling the stuff. Just thinking about that razor-sharp edge slicing into your finger—cringe! Maybe it's a result of the digital age—I've forgotten how to handle paper. I no longer have the patience to decipher those cryptic origami instructions, either. Then again, I am currently devising a plan to create temporary housing out of all the junk faxes I get. Stay tuned.

1

SHOE-BOX ORGANIZER

INGREDIENTS

+ 6 empty shoeboxes
+ 2 plastic hangers

TOOLS

+ Ruler
+ Utility knife
+ Duct tape

01 Cut an opening out of one end of the shoe boxes, approximately ¾" from the edges.

02 Open the lid to one of the boxes and cut a slit in all four corner verticals of the box end so the plastic clothes hanger can stretch its legs out of the box.

03 Cut a slit large enough to fit the hook of the hanger through the topside long edge of the box lid, and then slip the top of the hanger through.

04 Stack your boxes neatly on top of each other, placing the box with the hanger on top.

05 Duct tape around all the boxes, close to both ends of the boxes.

06 Hang in your closet and organize.

2

MINI-HANDMADE BOOK

INGREDIENTS

+ One 8 1/2" by 11" piece of paper

TOOLS

+ X-Acto knife

01 Fold the paper in half the long way.
02 Open the paper.
03 Fold the paper in half the short way.
04 On one side, fold the edge of the paper back to meet the fold.
05 Flip the paper over and fold the edge of the paper back to meet the fold on the other side. Place the paper on the table so that you see a W when you look at the end. Cut along the center fold, stopping just before the cross fold (the paper remains in one piece). You're cutting through two layers of paper.
06 With your wrists above your fingers, hold the two halves from the top. Flip the paper on its side. You will have an open book with four sections.
07 Bring three of the sections together. Fold the last section on top of the other three so that you have a flat book.

You really should, you know.

3

PRINGLES MUG CADDY

INGREDIENTS

+ 6 Empty Pringles cans
+ 1 12" by 12" piece of 1/8" plywood
+ 1 sheet of decorative paper 12.5" by 12.5"
+ Handful of small finishing nails

TOOLS

+ Ruler
+ Pencil
+ Utility knife
+ Hammer
+ Glue gun

01 Measure and cut 5" up from the bottom of the 6 Pringles cans. Mark 1/2" down from the edge you just cut and trim at a diagonal.
02 Trace and cut the outer edge of the cans onto your decorative paper in two rows of three. Use this as your template to trace the cut circle forms onto your plywood.
03 On the cut edges of your cans, make 1/2" hash cuts about 3/4" apart. Hot glue gun the splayed tabs onto the plywood at the penciled marks.
04 Nail in a few finishing nails through a few of the tabs of each can.
05 Slip the decorative paper over the mounted cans and glue it down to the plywood.
06 Finish the edges by gluing down the excess paper around the edge of the plywood.
07 Mount to the wall and hang up your mugs.

Tarry no longer, chop chop!

Chapter: **02**

PLASTIC

MATERIAL OF THE FUTURE!

The Graduate's Benjamin Braddock knew plastic was the way to go, and he chose to follow Elaine to Berkeley instead. Fool! A little more than a century old, and look how plastic has taken off. The stuff is on fire! Just look around: Your phone, pen, computer, stereo, microwave, eyeglasses—even your underwear is composed of polymer blends. The story of plastic, very much like the story of paper, begins deep in the woods. Cellulose, a carbohydrate (the good kind) that forms the main constituent of the cell wall in most plants, is used in the manufacture of textiles, pharmaceuticals, explosives, and—yes—plastics. That's why we call food wrap *cellophane* (everything is so connected!).

Now jump from the forest to your average day in the concrete jungle, just doing what you usually do, meaning no harm. You start the morning with a to-go coffee on your way to work (plastic lid, possibly a foam cup), spend a few hours dodging e-mail or crafting clever, heavily edited responses, and presto—time for lunch! You're not enough of a planner to bring a Tupperware container full of leftovers every day, so you grab something at the sandwich shop nearby. With your salad or delicious grilled cheese (with a choice of slaw or macaroni) you are given a set of plastic utensils, a molded plastic container, and maybe the plastic-bottled soft drink of your choice, all of which is placed in a plastic bag and handed over to you with the *ka-ching* of the cash register—four disposable items that you will use once, over the course of thirty minutes, then toss in the trash, maybe even scoring a basket from where you sit. Nearly 30 percent of all plastics produced are used in packaging. Americans discard enough plastic utensils each year to serve the world a picnic every other month.

Americans discard enough plastic untensils to serve the world a picnic every other month.

Where does the other 70 percent of the pie go? Consumer products—such as cameras, utensils, and disposable razors—account for another 15 percent. The rest gets parceled out to automobiles, furniture, and electrical components, and other needy causes like hospital and military equipment. "Yes, but I recycle," you demur. Okay, but do you even know what the little number inside that triangle on the bottom of that cottage cheese container means? In many cities, collectors won't pick up anything with a number higher than two, and if a four or a five gets mixed in accidentally (a yogurt container, say), it can ruin a whole batch of slurry in the process. The sad truth, friend, is that less than 1 percent of all plastic gets recycled, and the portion that does can't be recycled again. Would you buy shoes you could only wear twice? We didn't think so. Worse, very few plastic containers can be reincarnated at all—even if they have a number surrounded by the three chasing arrows on the bottom. Many cities will only pick up #1 PETE and #2 HDPE. Our handy reference chart will tell you what the numbers mean (see page 49). Check with your city's recycling center to find out what they'll accept.

A WRAP ON THE HISTORY OF PLASTIC

START HERE

1862
Alexander Parkes, a British chemist noted for his work in metals, strays from the cold, sharp-edged stuff to develop an organic material derived from cellulose that when heated can be molded to any form, and retains its shape once cool.

1907
New York chemist Leo Baekeland creates the first all-synthetic plastic and calls it Bakelite. The substance forms an exact replica of the container it's poured into and won't burn, boil, melt, or dissolve in any acid or solvent. It's tough stuff, so the military uses it to make weapons and lightweight machinery during WW II. It's also used for decorative housewares (now sold on eBay).

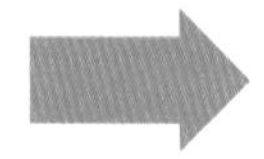

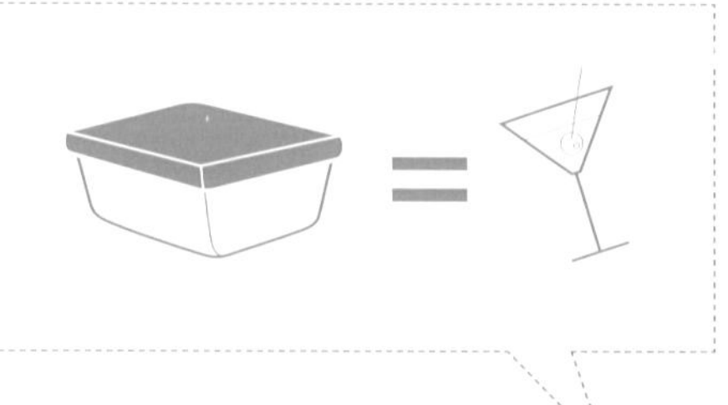

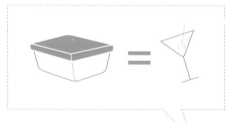

1933
Two British inventors cook up polyethylene, which, in its high-density form, will end up as the most popular plastic of all time (used for soda bottles, milk jugs, grocery bags, shampoo bottles, children's toys, even bulletproof vests).

1933
Hello, Saran Wrap!

1947
Earl Silas Tupper, a New Hampshire tree surgeon and plastics innovator, invents Tupperware. He threw the now infamous "Tupperware Parties" to market his invention directly to homemakers.

1948
Hello, Velcro!

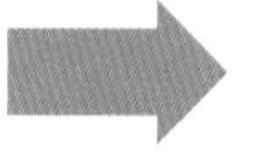

1959
Mattel introduces plastic Barbie doll. This marks a death knell for twentieth century feminism for nearly a decade.

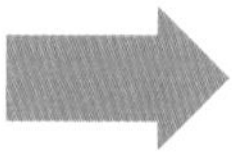

1966
Imagine life before plastic bags on a roll in grocery stores. I dare you! You cannot!

Not even two hundred years have passed

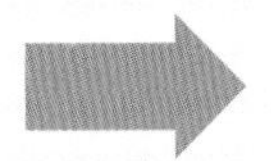

and we've advanced from moving

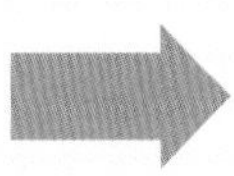

picture plastic to the artificial heart.

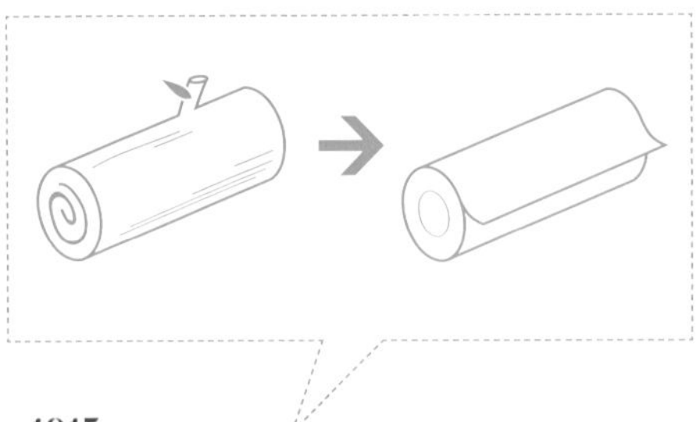

1866
Inventor John Wesley Hyatt, celebrated for inventing the billiard ball, knife sharpener, and domino, stumbles upon the chemical compound that makes up celluloid while searching for a substitute for ivory to make his billiard balls. Celluloid is the stuff of film for motion pictures. This marks the beginning of movie magic.

1913
Swiss chemist Dr. Jacques Edwin Brandenberger somehow manages to turn wood pulp (a form of cellulose) into the first fully flexible, transparent wrapping film. Having little imagination, he calls it cellophane.

1926
Searching for a PVC variation that would bind rubber to metal (read: make better tires), a BF Goodrich employee named Waldo Lonsbury Semon invents plasticized PVC, aka vinyl. Because it's cheap to make, durable, fire-resistant, and easily molded, it becomes a leather substitute for upholstery. A generation of kitsch furniture is born.

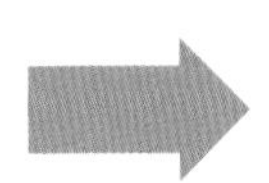

1937
Plastic is used as a casing to cover electrical appliances, like hair dryers.

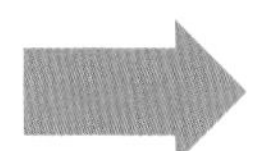

1940
DuPont Corporation introduces the nylon stocking. Women slink into them immediately and they become a runaway success.

1957
First Baggies and sandwich bags are introduced, creating a whole new school-lunch vibe.

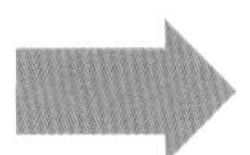

1957
Hello, Hula Hoop!

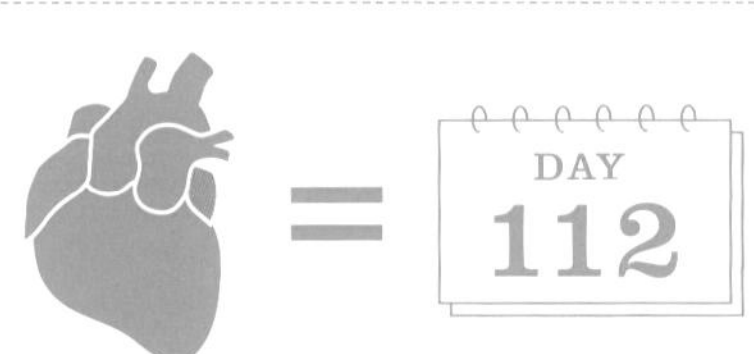

1977
The first salesclerk asks, "Paper or plastic?"

1979
The volume of plastic production surpasses the manufacture of steel in the United States just over a decade after *The Graduate* predicted it would.

1982
An implanted Jarvik-7 artificial heart sustains Barney Clark for 112 days.

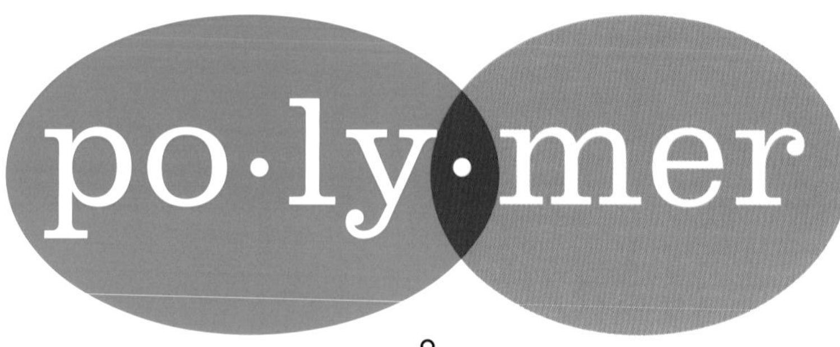

THERE ARE TWO TYPES OF POLYMER

Thermoplastics are the more popular of the two and make up the majority of products. Once formed, this polymer can be heated and remolded. It's easy to recycle, too, though it takes a not-insignificant amount of energy to do so.

Thermosets can't be re-melted. Don't even try. It's tough stuff

BEYOND TUPPERWARE

Learn about plastic, then buy some!

Go Polymers *(www.Go-Polymers.com)*
Buy and sell scrap plastic parts and virgin resin

Freecycle *(www.freecycle.com)*
Free supply of computer casings, water bottles, and other plastic paraphernalia

The polymers that make up plastic are really just a string of tiny elements—carbon, hydrogen, oxygen, and silicon—all linked together like a paper-clip chain. *Polymer* is a big word. Let's break it down. *Poly* means many, as in polyphonic (many sounds) or polygamy (much fun). *Mer* means "sea" in French, and though the idea of "many seas" is very poetic, here we're meant to take the Greek derivation: *meros*, or part.

An armchair psychologist would say that plastic is constrained by its upbringing and doomed to act like its parents. Only a total meltdown will make for a substantive change in shape and behavior. But change is possible. When heated, plastic will take virtually any form. Polymers are also lightweight and resistant to chemicals, and can be adjusted for varying degrees of strength (Saran Wrap does no one any harm, while a plastic air conditioner would flatten you if it fell from a window as you passed sweatily underneath).

PETE 1

» soft-drink bottles (lids off)
» dressing (ketchup, salad, peanut butter) bottles

» shampoo, dish- and laundry-detergent bottles
» flower pots

» fresh food packaging
» cable insulation
» water pipes
» garden hose
» cassette trays, jewel cases
» credit cards

» frozen-food bags
» soft honey bottle
» bread bags
» trash cans

» brooms and brush handles
» tissue holders
» white medicine bottles

» rulers
» license-plate frames
» foam packing
» egg cartons

Way back in 1988, when cell phones were as big as footballs and the home computer was a luxury item, the SPI (The Society of the Plastics Industry) invented a coding system, which assigns a number (between 1 and 7) to each consumer product to indicate the type of plastic resin the product is made from. The idea being, We can recycle your water bottle, but don't start throwing your Motorolas in the blue bins.

DUCHAMP'S CORNER

12 Alternate Uses for the Ubiquitous Plastic Water Bottle

Watering can
Chandelier (strung with X-mas lights)
Doghouse-roof tiles (flattened)
Decorative garden edge
Pencil holder (top lopped off)
Hand weights (filled with rocks)
Funnel (bottom lopped off)
Mini race-course markers
Cooler ice pack
Change holder
Terrarium (top lopped off)
Rain gauge

CD WALL MURAL

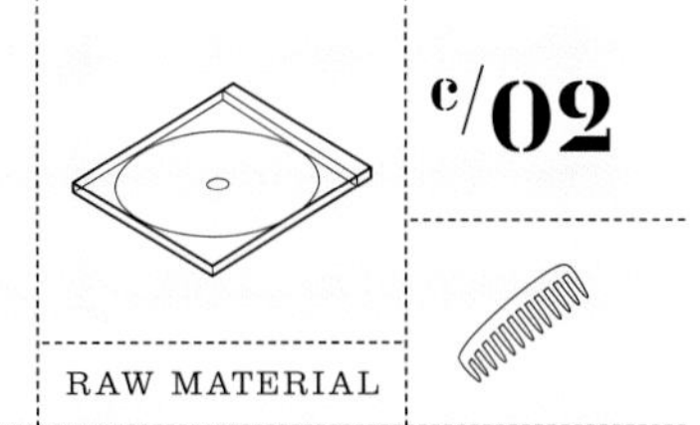

Have you switched to the soft pack of CDs? Are all your empty jewel cases starting to block the way to the kitchen? Time to make something from that mess of plastic brittle. Remember, jewel cases are fabricated from Thermoset, which can't be melted down and turned into two-liter Coke bottles. It's our way or the highway to the dump for these fellers. But look at all they have to offer: protection against the elements; translucency; clean, modern lines. For all those reasons and more, use your empties to make a wall mural. It's yet another step in your march against passive domesticity.

State that could be shrink-wrapped with the amount of low-density polyethylene plastic film that's made annually: Texas

CD WALL MURAL

INGREDIENTS

+ Large image of your choice
+ 55 CD jewel cases (4 cases for every square foot of image)
+ Adhesive Velcro (estimated use is 2" per jewel case you hang) or heavy-duty mounting tape

TOOLS

+ Goo Gone (optional)
+ Ruler
+ Marker
+ Utility knife
+ Scissors

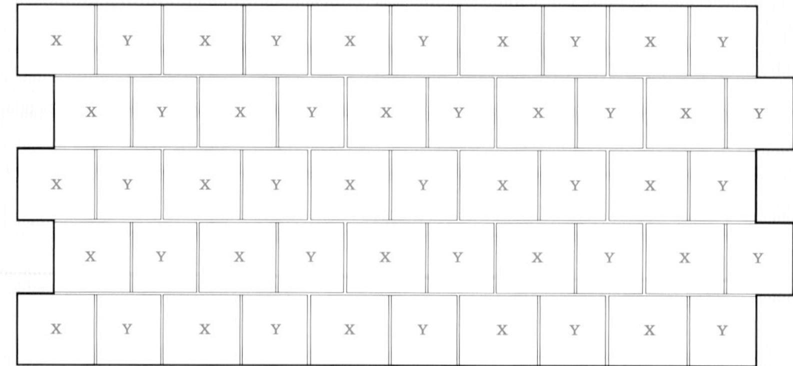

FIGURE A

1 If necessary, use Goo Gone to remove any residual stickers from the jewel cases.

2 Select a large image of your choice (we pixelated a landscape from one of Chris Ware's comics—thanks, Chris!—to make ours; see homokaasu.org/rasterbator).

3 Lay your image facedown.

4 Measure, mark and cut your image to fit the CD cases by first cutting it into horizontal strips of 4 5/8" tall (Figure A).

5 Cut alternating widths of 5 3/8" (for the back cover, panel X) and 4 7/8" (for the front cover, panel Y) as shown in Figure B. Make sure your rows layer like a brick wall, as that is how the cases will be affixed to the wall (Figure A).

6 Label the pieces as you go so you can keep them in order.

7 In order, insert the cut pieces into the CD cases.

8 Apply three squares of Velcro (with the male and female sides still stuck together) to the top, midsection,

Every day, he ate a frozen food called Tasty Bites that was delivered to him by the twelve-pack from an Indian mail-order catalog. The single serving packs of curry and rice came in a microwavable plastic container which, once emptied and scrubbed clean, he filled with dollar bills and sent home to Punjab.

Percent of average grocery bill used to pay for packaging (both paper and plastic): 10

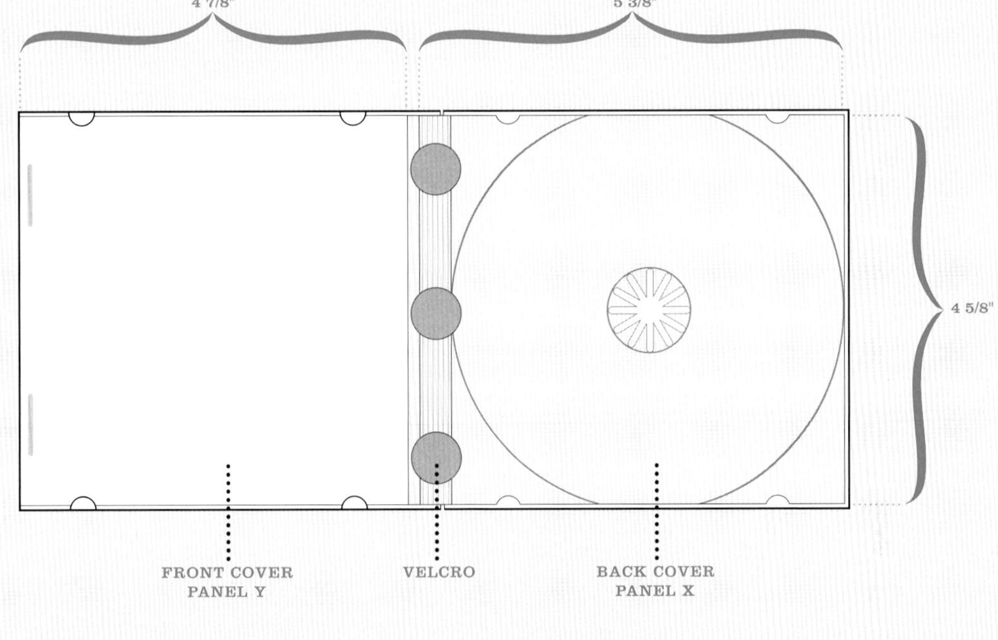

FIGURE B

and bottom of the wall-facing part of the CD spine (Figure B).

9 Remove the backing from the Velcro and affix your CDs to the wall to create your mural.

TIP: Choose a large image, and piece it together case by case. Creep it up a wall or outline a doorway. Because the final project is made up of smaller parts, you can put together a traditional square frame or form a partial shape that will make all who view it do the work of finishing the piece themselves.

Number of sea creatures killed every year by plastic bags and other plastic rubbish that's dumped into the ocean: **1,000,000**

TAKE-OUT CHANDELIER

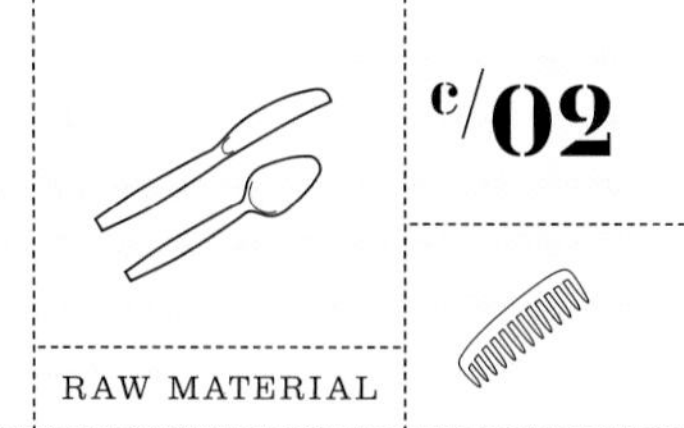

Choose one—Dangling crystal pendants are: a) too grown-up, b) too gala event, c) too expensive, or d) all of the above. You know the answer. This version, constructed from the plastic utensils you've been keeping around in case a picnic for forty crops up, makes the same grand statement without being precious. It's a nod to formality, but with a contemporary twist. What else would you do with the disposable flatware that every take-out order oversupplies? Plus, it's outdoor friendly—ideal for dressing up your porch light and keeping moths at bay with its armor of serrated knives.

Years it takes for plastic to degrade in landfill: **400**

What planet the foam cups made in one day would encircle if lined up: **Earth**

TAKE-OUT CHANDELIER

$25

INGREDIENTS

+ 2 thin, clear-plastic bowls, one large and one small (think flimsy party-store quality; the ones we used were 16 ounces, with a diameter of 7 1/4", and 24 ounces, with a diameter of 11 3/4")
+ 2 yards of fishing line
+ 100 clear plastic knives
+ 100 black plastic spoons
+ 35 clear plastic spoons
+ 250 paper clips
+ String of inexpensive Christmas lights

TOOLS

+ Cordless drill with 1/16" drill bit
+ Clear ruler
+ Fine-point Sharpie
+ Masking tape (to mark placement of holes)
+ Scissors
+ Hacksaw

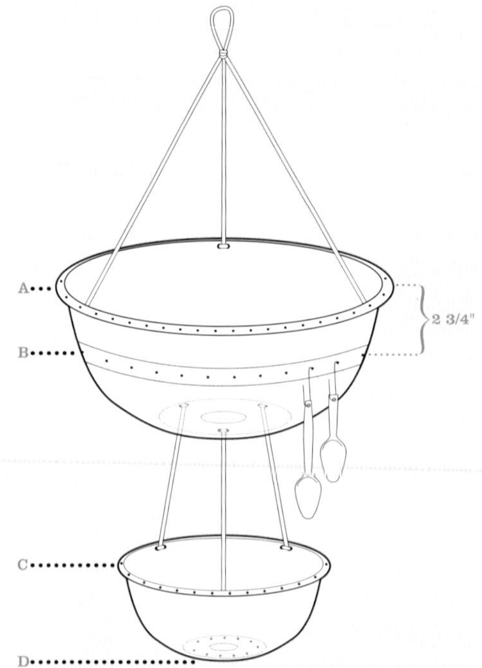

1 Drill holes into the larger bowl at the top rim, 1/2" apart (A).

2 Measure 2 3/4" from the top rim of the larger bowl and mark in four locations with your pen.

3 Lightly adhere your masking tape around the midsection of the bowl at the marks you just made.

4 Drill holes in this row, 3/4" apart (B). Remove tape.

5 Drill holes into the smaller bowl at the top rim, 3/4" apart (C).

6 Drill holes in the outer edge of the floor of the smaller bowl, 1/2" apart (D).

7 Drill three equidistant pairs of adjacent holes just above the row of holes (A) in the larger bowl. This is where the fishing line will hang to mount your chandelier.

8 Cut a small hole 2" in diameter at the bottom of the larger bowl. (This hole should be large enough to thread the lights through, but not so large that a portion of the lights cannot rest comfortably in the bowl.)

9 Drill three equidistant pairs of adjacent holes on the outer edge of the larger bowl's floor. This is where the fishing line will connect the large and small bowls.

10 Drill three equidistant pairs of adjacent holes at the top rim of the smaller bowl.

11 Cut a small hole, 1 1/2" in diameter, at the bottom of the smaller bowl.

Pieces of plastic debris that float on each square mile of ocean: **more than 46,000**

I had a retainer with a fake tooth attached to it. The plastic plate was so big, it filled my whole mouth. I had to slurp loudly just to swallow, which made things tough at school. But at Halloween, I enjoyed a sudden wave of popularity when I did the trick of taking it out and grinning fiendishly at girls. It scared their pants off.

Project No. 002

REMAKE THIS

WATER-BOTTLE LOUNGE

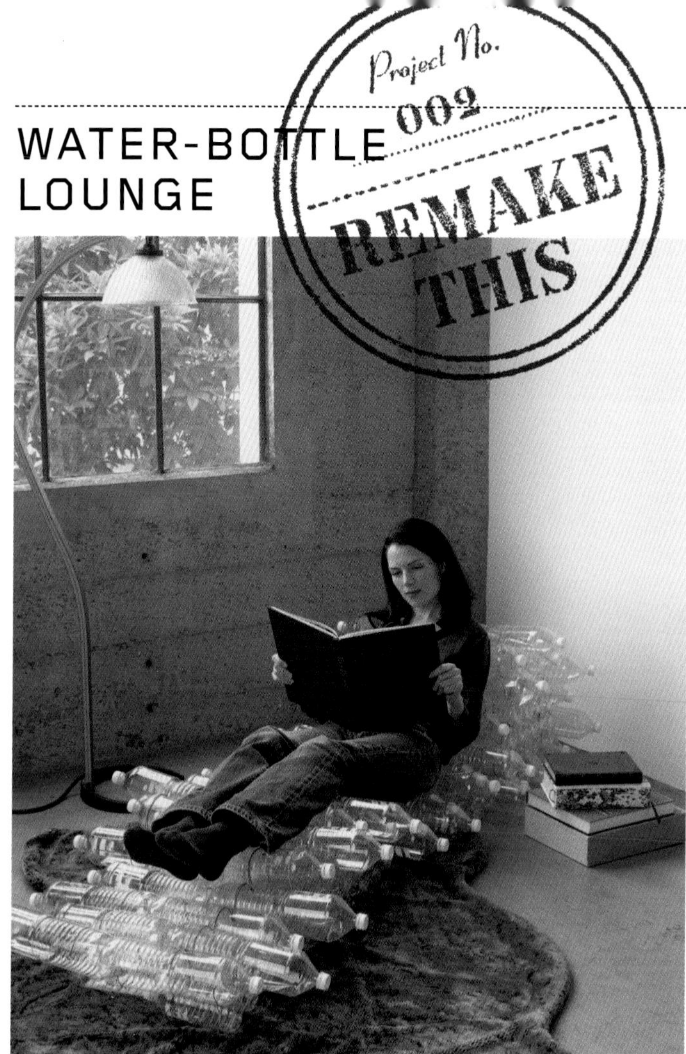

An empty water bottle lounge chair! We thrilled at the prospect. An ambitious undertaking, sure, but a great use for single-use plastic that requires a lot of energy to recycle. A little user testing dampened our enthusiasm, however. Know how it sounds when you crush a plastic bottle underfoot? Well, imagine the din produced by lying back on two hundred of them!

YES

1 Water bottles make good headrests.
2 We discard them by the locker-full on a daily basis.
3 We devised a binding system that would allow maximum surface area for your bum with the least number of bottles.
4 The design reminds us of the Louis Ghost Armchair by Philippe Starck, but much lower on the totem pole.

NO

1 Inspires fear in potential sitters (looks too fragile to hold weight).
2 Reclining makes thunderous noises.
3 The binding system is tricky at best, confounding at worst.

This hole will allow for the paper clip hooks to hang easily.

12 String the larger and smaller bowls together through the pairs of adjacent holes about 3 $^{1}/_{2}$" apart with the fishing line.

13 String the top of the larger bowl with about 9" of line, and tie together into one knot. See the diagram for illustration.

14 Determine how low you would like each row to hang and trim utensils to size with a hacksaw. (You can also adjust the hang height with the number of paper clips.)

15 Trim utensils as needed.

16 Drill holes into the utensils about $^{1}/_{4}$" from the bottom of the handles.

17 Hang utensils with paper clips. (In the chandelier pictured, each utensil hangs from one paper clip, with the exception of bottom tier of knives, each of which uses two paper clips.)

18 Thread the lights through the chandelier, carefully ensuring that a handful of the lights rest in a small pile in each of the bowls.

19 Mount to a deck overhang or ceiling and plug in for a faux-luxe glow.

NO MONEY? NO PROBLEM! FUNDING YOUR STARTUP ON UNSECURED DEBT IS THE AMERICAN WAY

The Six-Month Plastic Plan

Authorized Signature Not Valid Unless Signed

0% How to Start a Business on Credit Cards

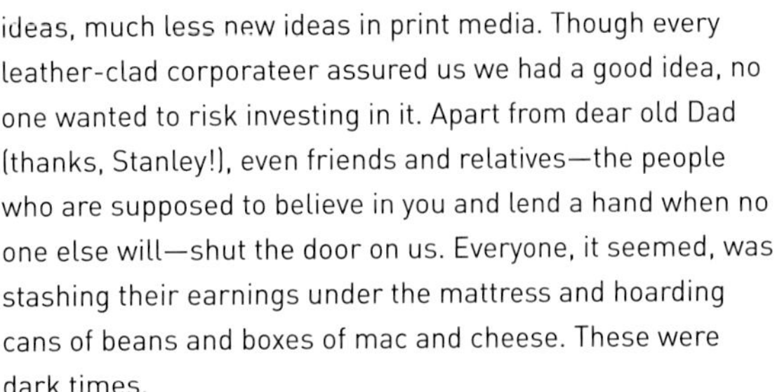

07/21/05	Pacific Bell	$ 45.05
08/15/05	Cash Advance	$ 500.00
08/17/05	United Airlines	$1000.00
08/26/05	Peete's Coffee	$ 50.00

Sure, you want to make your mark.

To not be yet another corporate schmo slumping in the conference room with after-lunch grease stains on your pants. "I have a Really Good Idea," you say shrilly, jutting your chin forward. More than that, your Good Idea could be a pioneering business, something that serves at once as a Force for Social Change and a Brash Cultural Movement that will motivate your generation out of its thick haze of ennui. They will cry out, "I want in! Hire me! Where do I sign?!"

There's just one little baby snag in your grand plan: You're broke, caught in the same consumer slugfest everyone else is in. Your meager wages get spooned out for rent, the Ikea–Urban Outfitters juggernaut, various media "needs" (magazines, CDs, DVDs), and a thirst for overpriced cocktails. Sure, sure, this is a reductive account of your much, much more complex and quietly striving life. But we mean to shake you up a little. Get inspired! Pluck up your ambition! We were in much the same overburdened, understimulated place when we launched a magazine three years ago.

We thought we had an idea that would:

1 → *Revive our creativity,*
2 → *Reduce lazy consumption,*
3 → *Address the zeitgeist,*
4 → *Lard our pockets with cash.*

Add to that heady mix an almost complete ignorance of how the business we were entering worked, lust for power, paralyzing insecurity, irrepressible belief in our concept, and barely enough in our bank account to buy a few computers and incorporate ourselves, and the sense of just how dangerous the first stages of any entrepreneurial sortie can be starts to show itself.

In July 2001, with a business plan and a prototype magazine in hand, we hit the pavement in search of startup financing. The dot-bomb stock-market crash was still bleeding in the road behind us. No one believed in new ideas, much less new ideas in print media. Though every leather-clad corporateer assured us we had a good idea, no one wanted to risk investing in it. Apart from dear old Dad (thanks, Stanley!), even friends and relatives—the people who are supposed to believe in you and lend a hand when no one else will—shut the door on us. Everyone, it seemed, was stashing their earnings under the mattress and hoarding cans of beans and boxes of mac and cheese. These were dark times.

Where do you turn when you can't cadge money out of anyone? To Uncle Sam! *That old goat must have a handout for young dreamers with a glint of Yankee opportunism in their eyes,* we thought. We started with federally financed programs endowed to subsidize women (we qualified) and minorities (we sort of qualified) in business, and we chatted up the Small Business Administration, which extends loans to those with collateral, such as property or a few cows. We weren't political enough or connected enough or seasoned enough to land the first two lenders on our short list, and to gain the support of the SBA we would have had to sign away our mothers' houses—the place we hung our first Duran Duran poster! Made out on the roof! Overboiled an egg! No! As desperate as we were to get our Obvious Media Empire off the ground, this was not part of the plan.

Though every leather-clad corporateer a$$ured u$ we had a good idea, no one wa$ taking ri$k$.

The only other place to turn was to those cornerstones of capitalism: Citibank, Bank of America, AT&T Universal, and American Express. Though anyone who grew up with an entrepreneurial bone in his body has been steeped in the bootstrapping myths and garage-inventor yarns that follow around captains of industry like Steve Jobs and Steve Wozniak of Apple, Fred Smith of FedEx, Ted Turner of CNN, and Martha Stewart of, er, Martha Stewart, it was still hard to fathom starting a magazine—the industry standard being that you should have no less than $5 million in the bank to launch—on scrappy savings and credit cards alone. Three years later, we're here to say, It Can Be Done! It requires some management finesse, eagle-eyed record-keeping, and the lionhearted flintiness of a mental-health worker, but it can be done.

Take anything you can get, Di$cover included: Thi$ i$ not the time to be fu$$y.

Here's how:

1. *Apply for a handful of credit cards.*

Draw up a projected budget for the first six months and acquire as many cards as it takes to add up to the credit-limit amount you'll need to purchase equipment, cover the first few rounds of COGS (cost of goods sold), and feed and water anyone whose brain you need to pick about the business you're entering or with whom you're trying to close a deal. Take anything you can get, Discover included: This is not the time to be fussy. At one point we had a total of ten cards from various institutions. Because we had decent credit when we started out, most of the lenders we called were falling over themselves to start us out at low introductory rates (between 5 and 14 percent annualized). If you have a personal credit card—which, these days, is like saying "if you have two lungs"—get started by transferring your balance to a new card to get in on the best rates. The trick with this stuff is to get lending institutions to offer you the lowest rates available for the longest duration possible. (When Melinda, your sunny customer service rep, comes back on the line to say, "We can offer you a 1.99 annual percentage rate until the transfer is paid off," it's like hearing every gospel choir in every church across America simultaneously break into song.) If you routinely pay off balances, or at least pay more than the minimum, and you keep spending, you're likely to get these incentives (in the form of courtesy checks or new card offers) frequently enough to start what we call the Great Plastic Shell Game. If you're a resident of the United States, you're eligible to play.

2. *Learn the rules of the Great Plastic Shell Game.*

Or, more biblically, rob Peter to pay Paul. The object of this game is to transfer something with negative value (i.e., debt) to gain something of positive value (i.e., lower interest rates) all the while attempting to extend the terms of paying off whatever you need to keep your Little Company That Could rolling for as long as you can, or at least until your next infusion of cash. So, first you make a payment on COGS or equipment or travel or power suits or deal-making dinners using card number one. Then you pay that balance, in full, thirty days later, using card number two (via electronic or check balance transfer), sidestepping any interest charges. Once credit card bill number two arrives in your mailbox, you handily pay off that balance in full, thirty days after it arrives, using card number three. You've now had a total of somewhere between sixty and ninety days to cough up the cash. This little song and dance can't go on indefinitely, of course—ultimately, you'll want to start paying off cards completely before an errant bill goes unpaid, resulting in steep late fees and a catastrophic bounce-up to the lender's highest percentage rate (24 percent and higher in some cases, with six months of parole before you prove to them you're not a felon or on your way to Mexico or something). But when you're starting out, this extra month or two grace period can make all the difference in terms of cash flow. Sometimes revenue just doesn't flow in the tidal way you imagined. Other times you're waiting for vendors to pay you and you don't want to ruin the relationship by pestering their accounting department. Once your little outfit settles into regular monthly revenue, you can think about trying for less-expensive financing, such as the now-more-confident-that-you-can-make-a-buck angel investor (best), traditional small-business loan (second best), or American Express card (a necessary evil that offers high credit limits but requires you to pay off the entire balance each month—ouch).

3. *Eventually, pay the piper.*

Suffer no illusions: Psychologically, the Great Plastic

Shell Game takes a toll greater than all the cumulative interest fees you rack up by the end of the year. Credit cards are incredibly expensive money. You will agonize over your debt. You will lie awake at night; your mind will curdle. By the end of our first year, we had $50,000 of unsecured revolving credit. If you think we had that many cows to cash in to pay it off, you've obviously never been on a farm. We had no cows—no homes even—and no rich uncles to airlift us out of this mess. Though we may sound blithe about this startup strategy, we paid the price in antianxiety therapies. The moral: Pay off this kind of debt as soon as the company makes any money at all. We know, you want to buy a bottle of bubbly, maybe expense yourself a sweet Aeron chair. Resist. Pay the plastic man.

Sure, you can hold on to a few cards to gird your business whenever cash runs short, but it's important to get rid of as many of them as possible as soon as possible. Why? You're still paying a premium for this money (especially if you're only ponying up the minimum, which is essentially paying interest and no principal on your loan—nibbling away at a bitter pill you know you'll eventually have to swallow). Currently, we've paid off all but two accounts—AmEx and a bank line of credit that covers our asses when our balance runs into the red. If you can't manage to repay the loan sharks, there's always Chapter 11, which isn't really the personal apocalypse it's made out to be. Sure, your credit rating will be in the toilet for seven years, but that just means you'll be using only cash and can finally get out from under the collections cronies who are threatening to go all Tony Soprano on you.

Postscript

Why is it okay to pull one over on the vaunted financial institutions that provide us with convenient machines that spit out money on every corner? Here's why: From the time you're in college, banks are trying to get you addicted to spending more than you have. Some kids complete their four-year cycle only to find themselves beneath a staggering mountain of debt (we've heard figures up to $60,000), never having learned how the illusion of free money actually works, or even how to manage a monthly budget. Americans have now borrowed more money than the federal government—we've outdone a shameful $5 trillion-plus deficit! Here's another reason: Credit-card companies take advantage of every gaffe you make. As we said before, at first opportunity they'll jack up your rates, slap on late and over-limit fees, and generally swindle you in ways that make it impossible to do anything more than accumulate interest. Not to mention the grim reality that they screw the less fortunate among us who need advances on cash just to get a hot meal. As Robert D. Manning, a professor at the Rochester Institute of Technology and an expert on the credit-card industry as it relates to poverty, told the Motley Fool, "the most economically disadvantaged or financially indebted [consumers] are increasingly relegated to the second tier of the financial services industry"—pawnshops, rent-to-own stores, payday lenders—"where interest rates typically range from 10 to 40 percent or more—per month!" This fastest-growing segment of the financial services sector invites the participation of the largest first-tier banks, like Wells Fargo and Bank of America. According to Manning, in 1997 Wells Fargo entered into a joint venture with Cash America, the largest pawnshop company in the United States, to develop a state-of-the-art system of automated, payday-loan kiosks. I mean, the nerve!

You will agonize over your debt. You will lie awake at night; your mind will curdle.

Anyone with a chestful of guile and a decent idea can launch a business, filling the capitalist aorta with fresh infusions of jobs, goods, and services. Remember, it's your duty in this land of milk and honey to be all that you can be. Why agonize over screwing the very system that's screwing you? If they're going to practice usury and hit you up with dodgy transaction fees, you might as well take them for all they're worth. We're all reaching for the same peach in the end: unlimited profit, increased sexual magnetism, and a new iPod.

Support Education

NO-SEW MESSENGER BAG

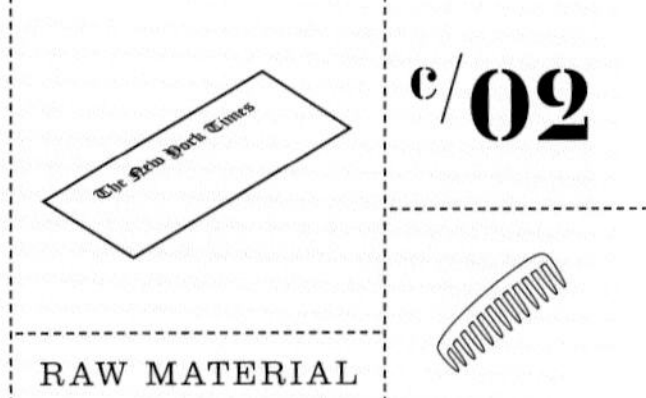

The narrow plastic tubes that our daily news gets stuffed into each morning do the indispensable job of bringing the world to our doorstep. Often doubled, the bags work like a skydiver's jumpsuit, protecting the paper from the wiry shrubs and other driveway danger zones into which it's cast. To accomplish this they must be durable, both tear- and waterproof—in other words, ideal material for a messenger bag. When several layers of the plastic are ironed together, they form thick, tarplike patches that can be fused with heat—no sewing required. Melt your way to your own wearable bag.

Millions of plastic bottles Americans use every hour: 4

Number of plastic bottles recycled in the United States: 1 in 4

NO-SEW MESSENGER BAG

$10

INGREDIENTS

+ Wax paper
+ One month's subscription worth of bags (25 to 30 in all—we used the large blue *New York Times* Sunday edition bags)
+ 1 yard of black webbing
+ 2" side-release buckle
+ 2" Ladderloc adjustment buckle
+ 2 pieces of 2" by 2" Velcro

TOOLS

+ Iron
+ Ironing board
+ Tape measure
+ Scissors
+ Sharpie

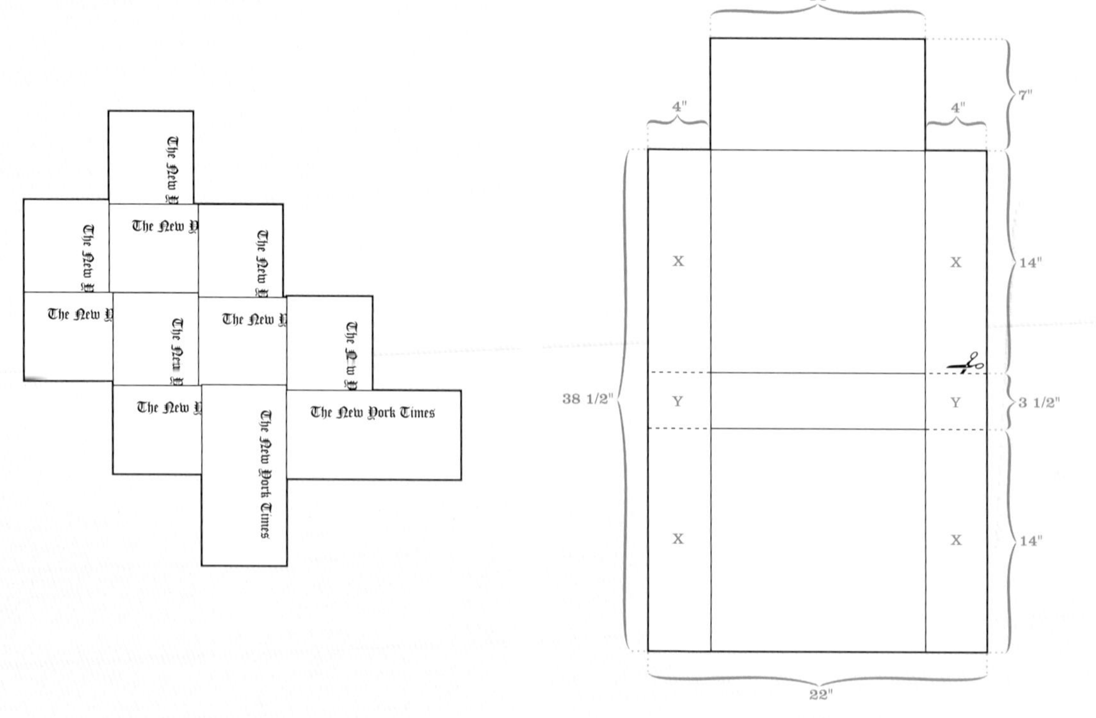

FIGURE A

FIGURE B

1 Lay down a large sheet of wax paper on the ironing board and arrange newspaper bags together in an alternating, interlocking pattern (Figure A). Cover the bags with another sheet of wax paper.

2 Iron on medium setting over wax paper.

3 Quickly peel back wax paper while the plastic is cooling.

4 Keep adding plastic bags (remember to sandwich them in wax paper—you don't want to ruin that iron!) until you have made a sheet about 2½' by 4'.

5 Pick which side is your display side and iron a few more layers onto the inside of the 2½' by 4' sheet. Your

Number of recycled two-liter bottles it takes to make the fiber fill for a ski jacket: 5

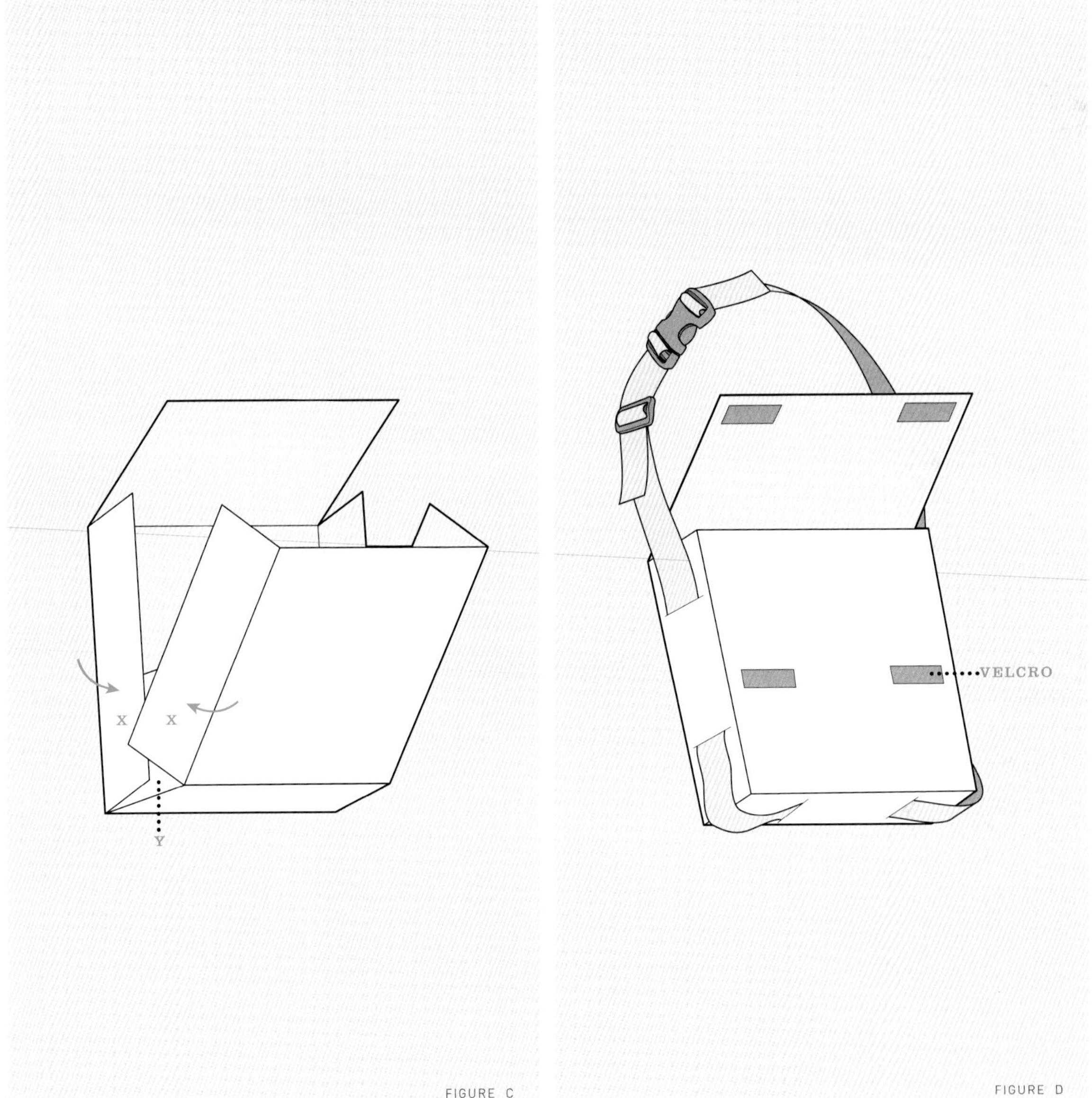

FIGURE C

FIGURE D

sheet should have the approximate thickness of an all-purpose tarp.

6 Measure and cut oul the perimeter of Figure B.

7 Iron the X-marked panels on both sides. This will form the short-width walls of your bag (Figure C).

8 Close off the holes on the outer bottom corners by ironing Y panels to the inside of the newly ironed X walls.

9 Cut 2" centered slits 2" from each bottom and top edge (Figure D).

10 Thread webbing through the slits (Figure D).

11 Attach buckles.

12 Affix Velcro to the inner lid and front for easy closure.

13 Load up and jet.

"Just one word:

Plastics."

Photograph © Bob Willoughby, 1967

Tide
BLEACH

ULTRACLEAN COATRACK

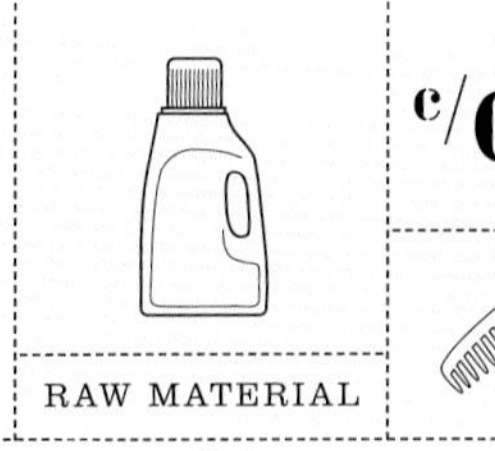

c/02

There are few things less imaginative than the coatrack. Its purpose lies solely in its function, leaving just two options when it comes to its form—a vertical rod, or pegs jutting from the wall. Mid-century-design power couple Charles and Ray Eames invented a more playful version: a wire frame bedecked with cheery wooden balls. This detergent-bottle hanger was inspired by that blast of color, with slanted handles that resemble the hooks of the Eames original. Suspended from your ceiling, it looks like sculpture when empty, and saves you floor space. Bonus: Everything smells laundry-fresh.

I grew up in Buckhead, GA, where suffering through scorchers was a rite of passage. My friends and I had a ritual we called the "hot game." We'd cover our bodies in bubble wrap and sit in the attic. When we couldn't take it anymore, we'd run downstairs, strip, and shiver in front of the air conditioner. That was our kind of fun.

ULTRACLEAN COATRACK

$8

INGREDIENTS

- 6 small empty detergent bottles with stout handles and caps
- 6 wine corks
- 6 ½" plumbing pipe plastic end caps (these were free)
- Extra detergent bottle cap
- 4' vinyl-covered clothesline
- Eyebolt

TOOLS

- Fine-point Sharpie
- Utility knife
- Cordless drill with ½" spade bit
- Linesman pliers
- Epoxy glue or other adhesive

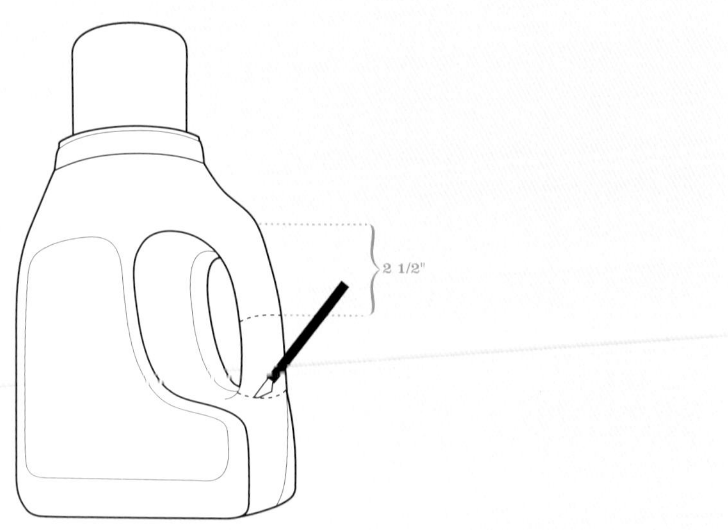

1 Rinse and dry your detergent bottles.

2 Mark the height of the hook you wish to create out of the handle (ours measure 2 ½" down from the top).

3 Using your utility knife, carefully cut the place you marked and the base of the handle to create the hook.

4 Cut a wine cork in half across the girth and whittle to size in order to fit into handle hole. This will help close off the handle opening and provide the necessary rigidity.

5 Repeat steps 2 through 4 for the rest of the bottles.

6 Glue on the ½" plastic end caps with your adhesive. Let dry overnight.

7 Drill a ½" wide hole through the center of each detergent bottle cap.

8 Drill a ½" wide hole through the bottom of each detergent bottle, aligned right above the hole in the cap.

9 Thread the laundry line through the top of the single cap and knot onto the eyebolt. This is how you will secure your coatrack into the ceiling.

10 Thread the rest of the line through your bottles, cap sides down.

11 Fashion a knot to secure the line from slipping through the ½" drilled hole.

12 Install by suspending from your ceiling.

13 Take it off and hang it up.

TIP: For extra bling, you can thread an inexpensive rope light through the bottles to brighten up a dark entryway.

EXTREME MAKEOVER BE DAMNED!

FOLLOW OUR 100 PERCENT GUARANTEED, NO-KNIFE PLAN

How to AVOID Plastic Surgery

Is that built-up cartilege? How unladylike!

Oh my, I think I spy some lines on your forehead . . .

It must be tough carrying around those bags. Mercy!

Those cheeks are sagging, sweetie!

Has your nose gotten bigger . . . from too much liquor?

Is that a triple chin?

Sure it's a stretch,

but you'll thank us for this one. In ten years, when your neck looks like a turkey wattle, you'll come back to this chapter and ask yourself, *Why didn't I follow directions? Why did I not believe?* O ye of little faith, listen up: Even if you have perfectly self-hydrating, alabaster skin with nary a line; even if you've worn a hat and sunglasses since you were five; even if you've never known a moment of stress, your face will drop—gravity is nature's cruelest law. It hangs on your facial muscles like a thousand possums. Your jowls, however strong, cannot hold up to these marsupials of the air. Your features will slacken and lilt until the continents of face, neck, and shoulders melt into one.

What can you do to stop the drop?

→ Don't smile. Have you ever noticed how miserably lined happy people are? They may feel great, but they look like crap. If you're too polite not to do something, try nodding in agreement instead. If you're overcome with joy and can't help but show teeth, at least keep the grin in check. Try not to jimmy the corners of your lips into your cheeks or scrunch your eyes into slits—that's where the cracks in the firmament show first.

→ Don't think. Ponderousness causes brow furrowing, facial ticks, and squinting.

→ Exercise your face. As with any part of your body, you need to give your mug a regular workout. With some discipline (less thinking, less smiling) and a few daily reps, you can tone your facial musculature and project a vibrant, evergreen look. This simple regimen helps improve your skin, too. Dermal tissue gets worn out and indolent as it grows older, but with proper flexing it can be imbued with new vigor. Increased blood flow delivers nutrients and flushes out all the nastiness you pick up from the world around you. What a feeling!

→ To whip your features into shape, do this no-knife, non-plastic face lift daily for at least a month, or until you feel too silly to continue. Start young (35 years of age at the latest), before things go south for good.

For the dour-faced and rubber-necked:

1 → Tilt your head back and look at the ceiling. Start chewing slowly while keeping your lips closed (a stick of gum helps). You will feel the muscles working in your neck and throat. Chew at least twenty times.

2 → Repeat step one, but instead of chewing, pucker your lips as if you are trying to kiss something dangling above you (a floating boyfriend or girlfriend helps). Keep your lips puckered for a count of ten, then relax and lower your head. Repeat five times.

3 → Repeat step one, but again, instead of chewing, open your lips and stick out your tongue as if trying to touch your chin with the tip (think Gene Simmons). Keep your tongue in this position for a count of ten, then relax. Weirdly fun, no?

4 → Repeat step one, then move your lower lip over your top lip as far as you can and keep it there for a count of five. Repeat five times.

5 → Lie down on your bed, belly up and with your head hanging over the edge. Slowly bring your head up toward your torso and keep it there for a count of ten. Relax and lower your head again. Repeat five times. Do this in private.

6 → Sit upright. Open your mouth and jut your jaw forward. Hold for a count of ten, then return your mouth to the starting position. Repeat five times.

7 → Sit upright. Place your tongue along your bottom gumline and extend it outward. Be careful not to overstretch. This looks ridiculous, but it really helps. Repeat ten times.

For the bleary-eyed:

First, make sure you use a good moisturizing day cream with at least SPF 15, and a nourishing night cream to help smooth out the crow's feet and smile lines you create with all that politeness. Use the following set of exercises to deal with baggy, puffy eyes and droopy eyelids:

→ Tone your eyelid muscles by pressing two fingers on either side of your head, at the temples, while blinking rapidly. Repeat five times.

→ Sit upright with your eyes closed and relaxed. First look down (eyes closed, please!), and then look up as far as possible. Repeat ten times.

→ Sit upright with your eyes closed and relaxed. Lift your eyebrows and stretch your eyelids down. Farther! Keep in this position for a count of five. Repeat five times.

→ Sit upright, looking straight ahead with your eyes open. Look up, then down, keeping your head still. Repeat ten times. Then look left and right. Your eyes should really be feeling it now! Repeat ten times.

For the browbeaten:

The area on the forehead between the eyebrows is open-trench warfare on the skin. Here's how to help smooth out your battle scars:

→ Frown as much as possible, and try to bring your eyebrows over your eyes, pulling them in toward one another. Then raise your eyebrows as high as possible while bugging out your eyes. This will feel unnatural. Repeat five times.

→ Lie on your back on the bed with your head hanging over the edge. Raise your eyebrows as high as possible, eyes wide. You'll feel a little daft doing this, but just go with it. Repeat ten times.

For those of you who do not follow this regimen, here are a few terms you should familiarize yourself with:

Wattle → A brightly colored, fleshy growth on the throat region of a turkey. Turns bright red when the turkey is upset or during courtship. (See also: caruncle.)

Snood → A flap of skin that hangs over a turkey's beak. Turns bright red when the turkey is upset or during courtship.

Dewlap → A pendulous fold of loose skin hanging from the neck of certain animals.

c/02
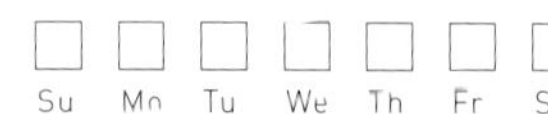
Su Mo Tu We Th Fr Sa

Why Don't I Do This Every Day?

TESTIMONIAL # 0035

NAME: **Casey Tilmanis**

FROM: **Berkeley, CA**

OCCUPATION: **Student, Triathlete, Aspiring Photographer**

Casey, why don't you do this every day?

Well, actually I do make things like this every day. As a college student on a limited budget, a little creative reuse can go a long way. Old jam jars become drinking glasses, mismatched socks become hand puppets, old shoes become flower pots . . . the possibilities are endless.

1

WATER-BOTTLE PLANTER WALL

INGREDIENTS	TOOLS
+ 12 1.5-liter AND .75-liter plastic bottles	+ Utility knife
+ 1/2" wood screws	+ Ruler
+ Old picture frame	+ Hammer
+ Piece of 1/8" plywood sized to your frame	+ Staple gun
+ Soil and plants	+ Power drill
+ Picture-hanging claw, wire, finishing nails	+ Scissors

01 Cut off the top 4" of a 1.5-liter bottle.

02 Cut away the front-facing half of your remaining water bottle, 4" down.

03 Repeat steps 1 and 2 for all the 1.5-liter bottles.

04 Cut away half of your .75-liter bottles and poke drain holes with a nail along the bottle rim.

05 Insert your cut .75-liter bottles into the cut 1.5-liter bottles.

06 Nail the sized piece of plywood to the backside of the frame.

07 Staple gun two rows of six bottles onto the plywood in an orderly fashion.

08 Fill each planter with 3 inches of soil.

09 Plant something.

10 Hang and water.

Why don't we?

2

SARAN-WRAP CHAIR

INGREDIENTS

+ An old metal chair frame
+ Shrink wrap (think industrial-sized plastic wrap)
+ Clear packing tape

TOOLS

+ Utility knife

01 Clean off the metal chair frame.

02 Wrap the plastic around the frame's backrest 10 to 20 times, depending on the opacity and tightness you want to achieve.

03 Cut the wrap with your knife and secure it with a small piece of packing tape.

04 Wrap 30 to 40 layers of plastic around the frame's seat for appropriate strength.

05 Cut the plastic with your knife and secure it with a small piece of packing tape.

06 Sit down.

We really should, you know.

3

COMB AND RULER MAIL SLOT

INGREDIENTS

+ 2 large combs of equal length
+ Wood yardstick

TOOLS

+ Handsaw
+ Sandpaper
+ Hot glue gun

01 Measure the combs and cut the yardstick into four pieces of the same length. Sand the edges.

02 Use the glue gun to apply glue along the length of the wood, lying flat, to the length of the comb's bottom edge.

03 Repeat and glue so each comb has wood glued on both sides.

04 Slot your mail in priority order.

Tarry no longer. Chop chop!

Chapter: 03

WOOD

THE GIVING TREE

Fancy cutting down all those lovely trees to make pulp or bloody newspapers and calling it civilization. —Winston Churchill, 1929

Animal bones make useful tools, but cavemen considering which materials to use for larger weekend building projects must have looked around for the tallest things in their neighborhood and thought, *Okay, we've got a couple of choices here: mountains or trees. Either we use our flints and chip off a bunch of rocks, or we take our bone knives and saw down one of these here trunks.* The decision was simplified considerably once the determination had been made that it was the fruit, and not the leaves or the bark, that was edible. (This was not obvious at first—why, for example, should wood have the property of being nutritious for termites but not for humans? It's very tricky, this case-by-case status of food.) Early man resorted to stone, as the anatomical shapeliness of fruit further suggested that trees might be living creatures that would actually bleed if cut. They spared the forests and carved their dream huts out of bedrock instead, with built-in firepits and wishbone hooks on the walls for hanging pelts.

After a while, though, the stone dugouts seemed a little chilly, crude even, and our pleistocene progenitors hankered for something warmer—the sun-dappled, curvy branches that hovered above. And so it came to pass: the first dark age of falling trees. What did they make out of the lumber bounty? Our forebears didn't really need walking sticks or shelves yet—hunting for the clan took up all their free time. So they whittled down their wood into aerodynamic spears, and rasped out bowls so the women could grind up buckwheat for pancakes.

Then, with the ancient Egyptians and Chinese, came more-practical things, like coffins, idols, and, eventually, chairs to raise themselves above the dirt. Nowadays, plastic has taken the lead as the cheaper, more pliable material of choice, but until this past century, most things started out in life as wooden blocks.

This makes more sense when you consider the primacy of wood in ancient thought. (By *ancient* we mean not long after Abraham roamed the Earth, around 2800–2700 BC.) The *I Ching, or Book of Changes,* a collection of sixty-four hexagrams that describes all of creation in terms of the interaction of yin and yang, describes five *Hsing,* or energy types, in the human soul: wood, fire, earth, metal, and water. According to the Great Teachers (Confucius—the guy who showed us the Way—wrote an introduction to the book around 400 BC), the wood element represents willingness and production—the new, the sudden, the ready-to-use. The life cycle flows from wood (willing) to fire (sensing): The wood nourishes the fire, and we gather around the flame as it heats the Earth. In the warm belly of the Earth exists metal (thinking), which can be liquefied, like water (feeling), and molded into tools for productive action (back to the wood element again). And so the cycle repeats. Wood, like the Duraflame of life, is our beginning and our end.

> Until this past century, most things started out in life as wooden blocks.

THE RINGED HISTORY OF WOOD

START HERE

400,000 BC
The spears from Schöningen (Germany) provide some of the first examples of wooden hunting gear. Flint tools were used for carving.

Bronze Age
Wood carvings include trees worked into coffins from Germany and Denmark and wooden folding chairs.

Iron Age
Fine examples of wooden animal statues (dog gods—there's a reason for the palindrome) are found at the site of Fellbach-Schmieden, Germany. Wooden idols of La Tène-date are known from a sanctuary at the source of the Seine, France.

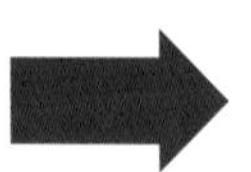

AD 100
The intricate glue-less and nail-less joinery for which Chinese furniture is so famous comes into fashion.

305
First wooden printing presses invented in China. And forget paper—they use it to carve symbols on a wooden block.

1927
First wooden roller coaster, the Cyclone, is erected in the United States. It was built for an initial investment of $100,000. And when it opened, a single ride was only 25 cents, and 35 cents on Sundays. Now for the stats: height: 85 feet, first drop: 85 feet, top speed: 60 mph (feels more like 100 mph), and length: 2,640 feet. It's still in operation.

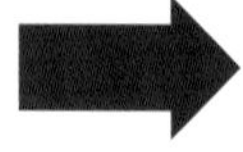

1933
First soapbox derby in Dayton, Ohio.

1940s
Charles and Ray Eames use shipbuilding techniques to mass-produce their now classic bent-plywood furniture.

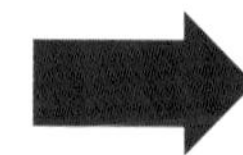

The ash bin of history.

From burning spears

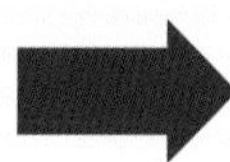

to the Duraflame log.

Neolithic times
Carved wooden vessels are dug up in places like the linearbandkeramic wells at Kückhofen and Eythra.

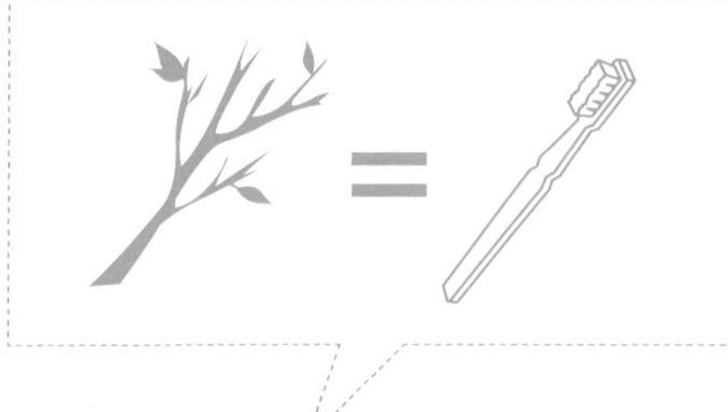

The long reign of Egyptian Pharaohs
Woodworking is depicted in many drawings. Some ancient Egyptian chairs have been preserved in tombs.

500
The original Buddha teaches his disciples to clean their teeth with a tufted toothbrush. In India they use twigs from the neem tree. In Sanskrit the toothbrushes were called *danta-kashuta* (*danta* means thirty-two, the number of teeth in our mouths, and *kashuta* means twig). There are no neem trees in China, so they use poplar instead.

600–700
Hello toothpicks! The spread of Buddhism throughout the Far East brings the Buddha's toothbrushes and picks to Japan from China and Korea. The Chinese and Japanese word for poplar is *yo* and for branch *ji*, hence the name *yoji*, or *tsumayoji*, for toothpick in Japanese.

1604
First wooden railway built in Wollaton, England.

1931
First wooden currency. When the Citizen's Bank of Tenino (say it: 10-9-0), Washington, defaulted during the great stock-market crash, the Chamber of Commerce decided to issue round wooden coins as scrip to use until they could fill the coffers back up with the real issue.

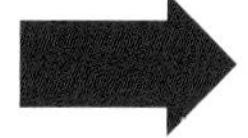

1952
The Boy Scouts of America hold the first pinewood derby.

1972
The Duraflame Corporation invents the self-igniting, paper-wrapped log.

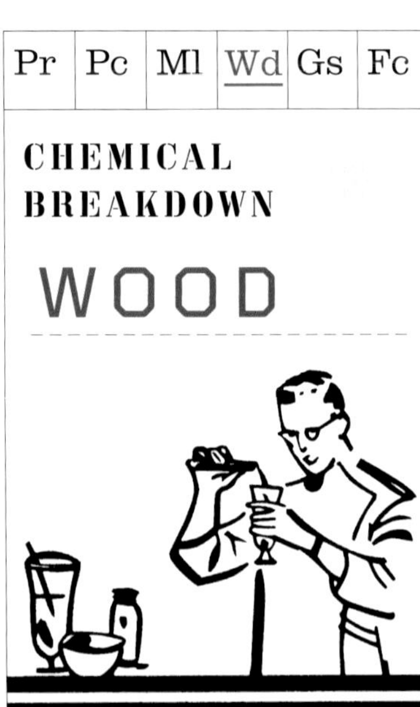

WOOD CHUCK

Where to score your cords

The American Tree Farm System
(www.treefarmsystem.org)
Educate yourself about renewable wood resources.

Old house parts
(www.oldhouseparts.net)
Naturally aged wood for home restoration or projects.

The Recycler's Exchange
(www.recycle.net/Wood/used/)
A reusable wood and lumber exchange.

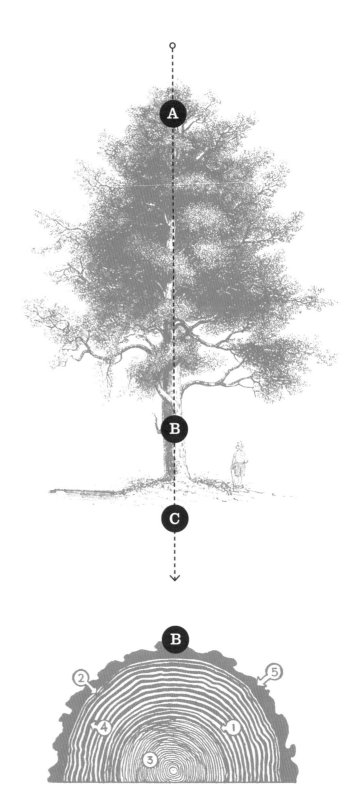

THE TREE OF LIFE

(Or, The Life of Tree)

Wood comes from trees. There is no other source. Trees, as we have said before in these pages, are made by the magic powers of the universe and contain the following organic compounds:

A// THE CROWN — Shade-giving leaves and branches

Anyone who passed high-school chemistry knows that leaves get their green color through a process called photosynthesis. But do you remember how photosynthesis works? (Don't disappoint the wide-eyed, questioning eight-year-olds of the world with a dodgy answer.) The sapling, made up of lignocellulose (hydrogen, oxygen, and carbon dioxide) converts the sun's energy into sugar and oxygen. Sugar feeds the tree (if not used immediately, it's stored in branches, trunk, and roots) and oxygen is released to feed us people.

B// TRUNK — The thing you climb

The tree's spine consists of five layers of wood that act as a gondola between the roots and the leaves:

1. Sapwood (the xylem layer): The youngest layer / These cells are the UPS crew of the trunk, delivering water and nutrients from the roots up to the leaves.

2. Phloem: The inner bark / Like sapwood, a conduit that carries sap and sugar from the leaves to the rest of the tree.

3. Heartwood: Main trunk support / Formed from the dead cells from the xylem layer (see next part). Heartwood is where the sugar, dyes, and oils are stored. They have the neat effect of turning the wood a dark sepia tone. Random fact: A vibration radiating out from the heartwood is commonly called heart shake.

4. Cambium: It grows / This very thin tissue produces new cells that look like rings, making the tree grow wider. Every season, the cambium helps to add a new layer of sapwood and xylem to the trunk.

5. Bark: The hard outer shell / Phloem cells, after a lifetime of deliveries, retire, shed outward, and become bark. This smooth or cratered outer layer protects the tree from insects, disease, storms, and extreme weather.

C// ROOTS — The subterranean part*

With its layer of tiny hairs, this finger-like substrate absorbs water and nutrients from the soil, stores sugar, and anchors the tree in place. The thickest lateral roots branch into thinner networks that extend as far out as the tree's branches, making it a mirror image of itself underground.

*Roots: The subterranean part (also cf. Haley, Alex, and early 1990s hip-hop)

CLOTHESPIN DOORMAT

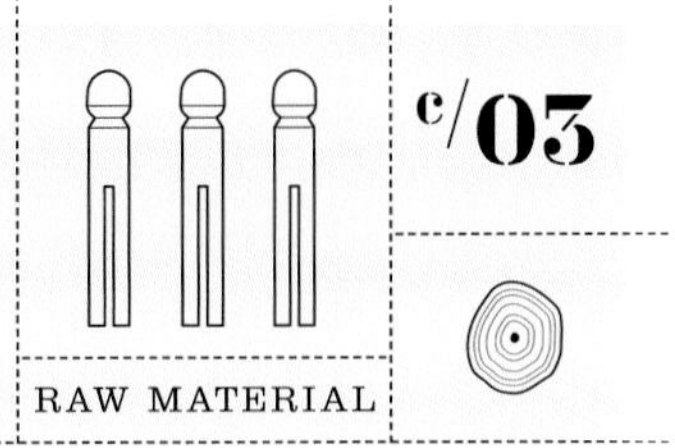

With their knobby heads and two-pronged appendages, these little wooden soldiers were once a homemaker's only line of defense against mildewed clothes. The advent of the electric clothes dryer chased them away, of course, and they can now be found in droves at salvage and thrift stores. In this variation on a theme, we've wired a flat of them together to take up a position just inside the entryway, after a more thorough scrubbing has been done on the mat outside. And when summer comes around, the stained pine cleans up good.

Where the tallest living Christmas tree (275 feet) is located: **Tasmania**

CLOTHESPIN DOORMAT

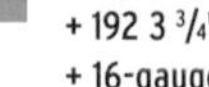

$20

INGREDIENTS

+ 192 3 3/4" doll-head clothespins
+ 16-gauge or narrower steel wire

TOOLS

+ Hand drill
+ Diagonal pliers
+ Drill bit larger than wire
+ Nose pliers
+ Hammer

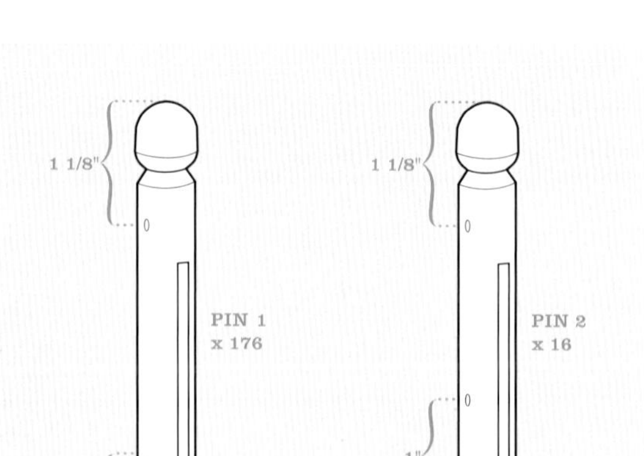

FIGURE A

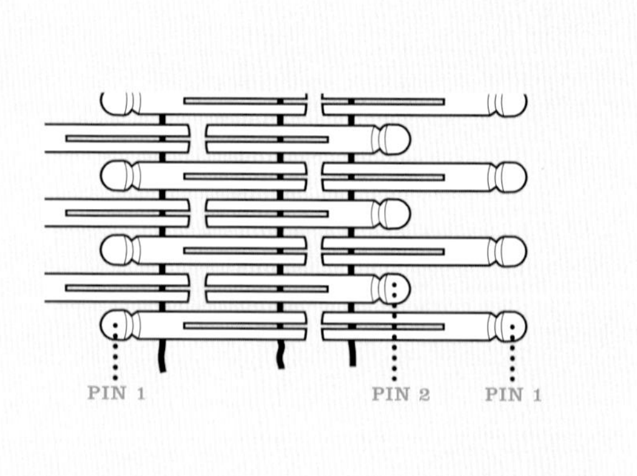

FIGURE C

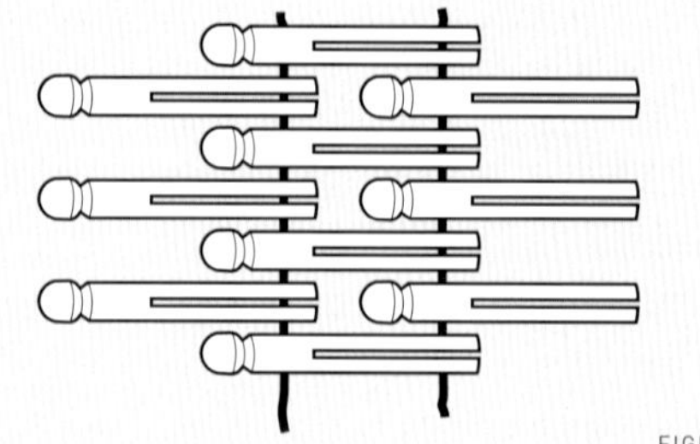

FIGURE B

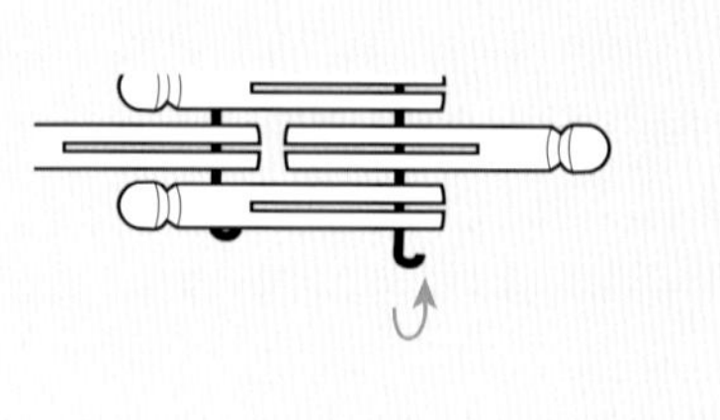

FIGURE D

1 Drill two holes through the 176 clothespins: one 1 1/8" from the head and one 1/2" from the foot (Pin 1, Figure A). Now drill two holes through 16 clothespins: one 1 1/8" from the head and one 1" from the foot (Pin 2, Figure A).

2 Use diagonal pliers to cut eleven pieces of wire, 27" long.

3 Sleeve the wire into the clothespins, alternating the height of each pin accordingly. (Be patient here! The wire will not slide through like butter. You will have to nudge the wire through a little at a time, sliding the upper connecting row over by hand every time you sleeve a new pin.) (Figure B)

4 On the penultimate row, you will flip the orientation of your pins so your outer edges have a continuous look. Use Pin 2 (Figure A) to make the penultimate row (Figure C).

5 Trim the excess wire to 3/4".

6 Drill a hole immediately adjacent to where the wire exits.

7 Use the pliers and hammer to bend, tuck, and hammer the excess wire into the adjacent hole (Figure D).

8 Flatten the mat by bending the web of pins into place by hand.

9 Set by the door and wipe up.

Number of trees it took to print the latest Harry Potter book: **about 110,000**

this is not a project

c/03 | WOOD

READ ALL ABOUT THIS DRAMATIC, LONG-STANDING, STILL-UNRESOLVED RIVALRY

the Challenger

CHOP STICKS

"ONCE YOU LEARN YOU'LL NEVER GO BACK"

— VS —

Reigning Champion

THE FORK

"THE DARLING OF THE WESTERN WORLD"

THE ONGOING UTENSIL WAR

WHY CHOPSTICKS ARE BETTER THAN FORKS: AN ARGUMENT AGAINST DISPOSABILITY (AND USER'S GUIDE TO PROPER ETIQUETTE)

FIRST, CHOPSTICKS ARE OLDER.

Though an exact date cannot be put on the first time a young Chinese girl pulled twigs off a branch and used them to pluck food from a large firepot, it is thought to have been around five thousand years ago (not long before Noah started building his ark!). In order to conserve fuel, early Chinese cooks chopped their ingredients into itty-bitty pieces so that they would cook more quickly. The tradition continues today—you don't see many whole chickens or ducks or slabs of beef coming out of the Mandarin Garden kitchen. What better tool to pick up each cubed morsel than a pair of thin splints? Chinese chopsticks are called *kuaizi*, which means "quick little fellows." Quick little fellows!

After years of slicing open their lips on the nubs that lined the twigs, ancient cooks eventually winnowed them down into two smooth, slender sticks. By then (around the 5th century BC) the rice sweepers found an ally in that gentle giant Confucius (him again), who as a vegetarian and animal lover believed knives to be savage reminders of slaughter—far too uncouth for the dinner table.

In his book on semiotics, *Empire of Signs*, the French scholar Roland Barthes argued that chopsticks suggested a cultural divergence that went beyond mere manners. He described the delicate movement of chopsticks as "maternal" and opposed to the "predatory" instinct of Western cutlery. "The chopsticks are the converse of our knife: they are the alimentary instrument which refuses to cut, to pierce, to mutilate, to rip," he wrote. "By chopsticks, food becomes no longer a prey to which one does violence, but a substance harmoniously transferred."

Those seeking a more hermeneutical explanation might pair the decline of the knife with that of the feudal warrior class, and the rise of the chopstick with that of the scholarly gentry. David Graeber points out that the intellectual elites brought with them different manners, inspired by Confucius (again!), whose vegetarianism led him to believe—long before any French philosopher alighted on the idea—that knives encouraged men in the direction of the barbarism of the slaughterhouse.

At European tables, diners used knives and spoons (when fingers wouldn't do). Because metalworking was expensive, Europeans brought their own cutlery to public mess halls—usually their personal daggers. With ready weapons and sufficient quantities of mead, violent outbursts were common.

Once Confucius had blessed the use of chopsticks, they spread from China to neighboring countries such as Korea, Vietnam, and Japan. In Japanese chopsticks are called *hashi*, which means bridge. We do not know the significance of this. (Perhaps it referred to the one thing the Japanese and the Chinese could agree upon.) The Japanese, fond of eating with their hands, originally used the implements to wave around in religious ceremonies, like wands. But the sticks caught on, and Japan now imports from China 96 percent of the 25.7 billion single-use "bridges" its citizens throw away each year. China itself discards more than 45 billion pairs, or 25 million trees' worth. Once upon a time the ruling classes refused to eat with chopsticks made of anything but silver, since it was believed that the metal would turn black upon contact with poison. If only the superstition had endured, the nation would be in much better shape environmentally. In 2001, environmentalists warned that if the current rate of timber use for chopstick production continues, the mainland will clear-cut its remaining forests in about a decade.

Any attempt to phase out or even limit chopstick use is bound to meet with strong emotional opposition. The utensils are believed to improve memory, increase finger dexterity, and even allow us to pay more attention to our food. Don't laugh. Eating with a fork is mindless: Just stab and lift. It can be done while reading or watching TV. But if you're picking up bits of food with wooden tongs, you have to pay more attention as the food moves from bowl to mouth. You have to oversee the progress, which brings a kind of mindfulness that's missing from the Western meal.

DROPPING YOUR CHOPSTICKS IS A SIGN OF BAD LUCK TO COME.

Ironic, then, that the English *chopstick* should have its roots in a perversion of the Chinese "quick little fellows," by nineteenth-century British traders and journeymen, who replaced that translated phrase with their own slang for "hurry up"—"chop chop!"

ETIQUETTE

Chopsticks should be held in the right hand, even by (smarter, but more endangered by mishap) left-handers. In East Asia, as in Muslim nations, the left hand is designated for the toilet, the right hand for eating. Though left-handed eating has become more common (many restaurants make antibacterial soap available in their bathrooms), ancient Chinese traditions die hard.

The sticks should encounter minimal contact with the mouth. It is poor table manners to suck on the tip of chopsticks.

Use serving spoons to get the food to your bowl before using your chopsticks. (Note, however, that in China it's not unusual to use your own pair to serve yourself, an often alarming custom to those not familiar with it.)

After you have picked up an item, it belongs to you. No backs.

Never rest chopsticks by standing them up in a bowl of rice. It's lazy, for one, and it's considered morbid, as it resembles a funereal tradition in which food is offered to the deceased.

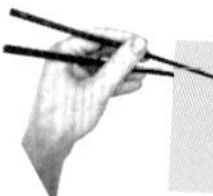

THE CHINESE WAY

Dishes are usually prepared to consist of bite-sized, pieces. If an item is too big or too small to be picked up, it is not intended to be eaten with chopsticks, but rather with a spoon. So logical!

The rice bowl is raised to the mouth, and the rice shoveled in using the chopsticks as a kind of forklift (bad pun, sorry). If rice is served on a plate, as is more common in the West, it is acceptable to negotiate it with a fork or spoon. Tweezing the rice grain by grain, for the chopstick-challenged Westerner, is akin to playing Operation with your food.

THE JAPANESE WAY

Having got all that wand waving out of their system early on, the Japanese decided that sticks should be used for eating and nothing else.

1. Do not point or gesture with chopsticks, and do not use them as drumsticks, no matter how hooky the song.
2. Do not dig around in the bowl for choice bits of food. Eat from the top and identify what you're going for before digging in.
3. Never stab or pierce food with chopsticks.
4. Do not move dishes around with chopsticks. They are not meant to do heavy lifting.
5. Do not lick or suck the ends of chopsticks—it's rude.
6. Do not let food drop off the ends of chopsticks.
7. Do not shovel food into your mouth with chopsticks like the Chinese. In Japan, no vessels, with the exception of soup or tea bowls, are raised to the mouth.
8. Never use chopsticks to transfer food to someone else's chopsticks, plate, or bowl. How unsanitary!
9. Place the pointed ends of the utensils on a chopstick rest when they are not in use.

THE KOREAN WAY

The thin, slippery surface of the metal chopsticks commonly used in Korea makes speed eating challenging, so Koreans tend to use a spoon for their rice and soup, and chopsticks for everything else.

THE VIETNAMESE WAY

Though they battled the Chinese for decades, the Vietnamese eat the same way—lifting the rice bowl to the mouth and shoveling the food in with the sticks. Vietnamese rice is very sticky, making this a tidier ordeal than it might sound.

PALLET BIKE RACK

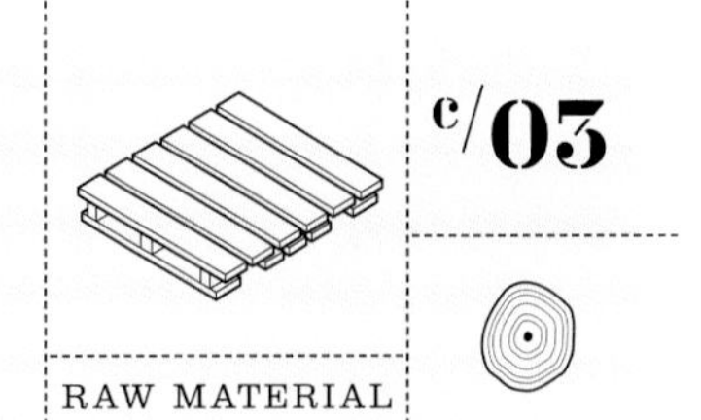

Ah, the humble pallet, keeping the cargo of our great consumer nation off the ground (and getting slimed and stained in the process). These old salts have seen many a port, and don't deserve to be abandoned at the road-side—especially when they make such sturdy bike racks! Set on end, the slats are thin enough to hold a wheel in place (you may have to do a little sanding or trimming for wider rims). Just prop it up with shelf angles, sand and paint for a more finished look, and give it a permanent place of residence in the garage.

How many people the oxygen produced annually by one acre of trees will support: 18

PALLET BIKE RACK

$15

INGREDIENTS

+ Small wood pallet
+ 1/2" wood nails
+ 4 10" by 12" shelving angles
+ 3/4" screws
+ Gray paint
+ Red paint

TOOLS

+ Tape measure
+ Pencil or marker
+ Hand circular saw
+ Jigsaw
+ Hammer
+ Sandpaper
+ Hand drill with Phillips head
+ Paintbrush

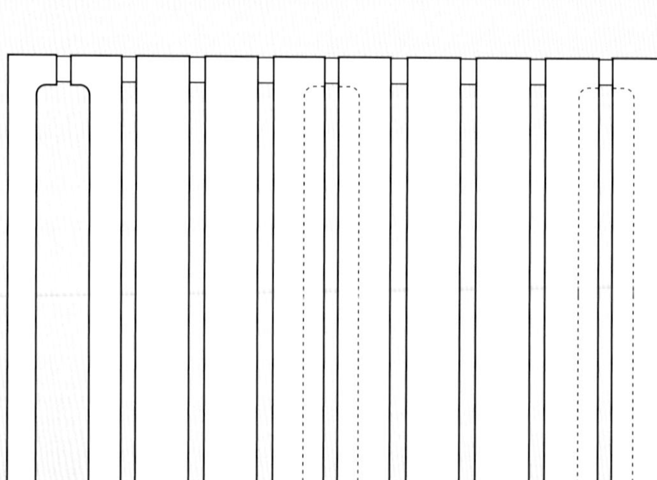

FIGURE A

1 Find a small wood pallet that does not have a horizontal middle bar on the backside for support.

2 Take note of the number of openings between the slats to determine which ones you'll be parking your bikes in. (Our pallet's slats were spaced approximately 1 1/2" to 2" apart.)

3 Lay the pallet front faceup to measure and mark what you are going to cut from your selected opening. It should be approximately 2 3/4" wide (Figure A).

4 Use your hand circular saw to trim the straightaway of the pallet slats on each side.

5 Use the jigsaw to cut a rounded

Number of stories high the world's tallest living tree stands: 37

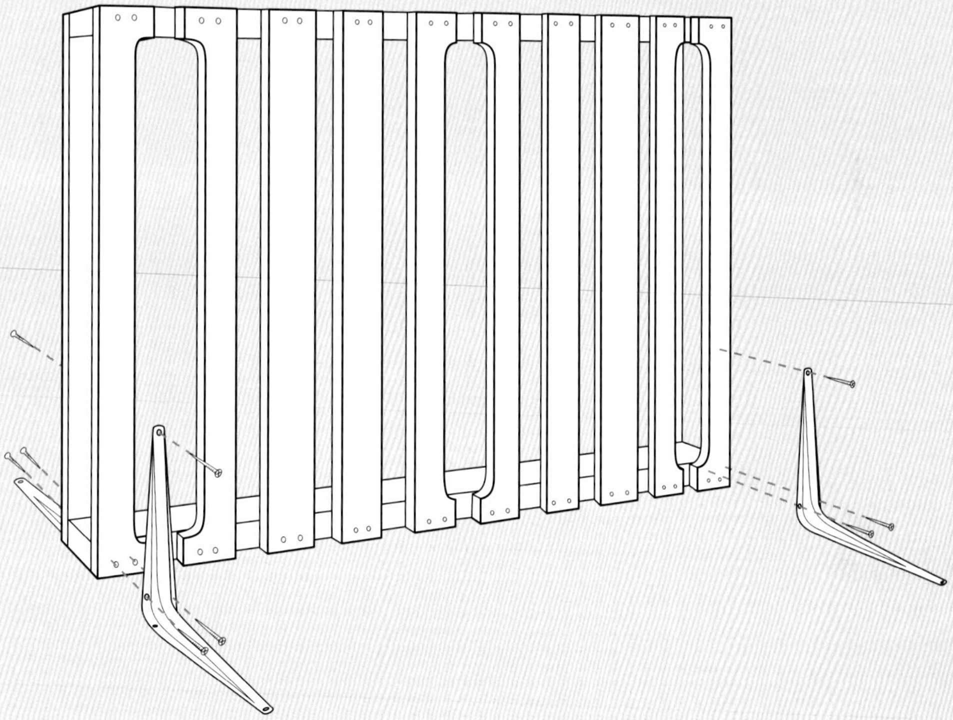

FIGURE B

finish for the tops and bottoms of your openings.

6 Hammer a couple of wood nails into the top and bottom of any loose slats that come unhinged from the pallet when you cut it.

7 Pallet wood is typically rough. You won't be able to sand it to a smooth finish, but lightly sand away any stiff splinters that could prick your tires.

8 Stand the pallet up and use the hand drill to screw in the shelving angles to the pallet on the outer edges of the front and backside (Figure B).

9 Paint the opening slats gray, with a red accent at the opening.

10 Let dry and park your bikes.

Number of trees per person it takes Americans to use up the CO_2 produced by the factories, engines, and vehicles that supply us with our daily necessities: 78

SHELLS & CORALS
BUTTERFLIES
TATTOOS
DDR DESIGN
TIKI STYLE
TIFFANY
HULA
TASCHEN
PEEPSHOW
THE CRITERION COLLECTION
8½

EAMES-STYLE DRAWER UNIT

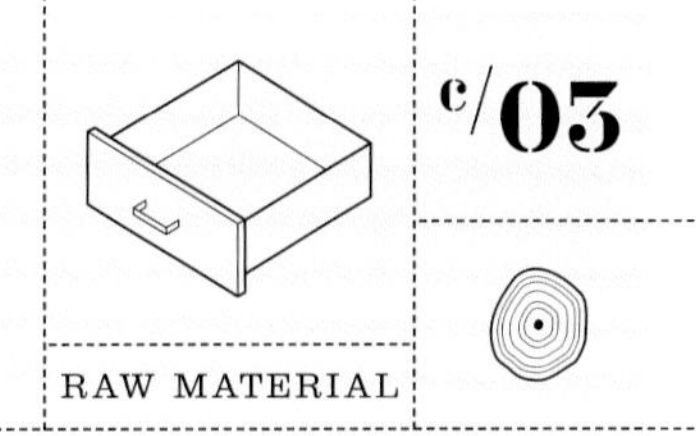

This is our answer to the cinder-block-and-plywood bookshelves ubiquitous in college dorms when we were in school. (Are kids still building those, or have they been killed off by the three-headed Ikea-Target-Wal-Mart monster? We need an update.) Inspired by the tower of stacked drawers by the Dutch collective Droog, and by Charles and Ray Eames's plywood storage units, we created this assembly of aluminum brackets and old drawers. Next time there's a big refuse day in your neighborhood, bogart some of the more handsome dressers left at the curb and build your own.

Tons of pollution removed each year by an average mature tree: 20

EAMES-STYLE DRAWER UNIT

$60

INGREDIENTS

- + 3 cast-off drawers
- + 4 8' aluminum angles, $^1/_8$" thick
- + Handful of #8 $^3/_4$" flathead Phillips screws
- + A few 1 $^1/_2$" carpenter's nails
- + Black semigloss paint (optional)

TOOLS

- + Tape measure
- + Permanent marker
- + Hacksaw
- + Sandpaper
- + Hand drill
- + $^1/_8$" drill bit (for the starter hole)
- + $^1/_4$" drill bit (for the sink part of the hole)
- + Phillips-head bit
- + Hammer
- + Paintbrush (optional)

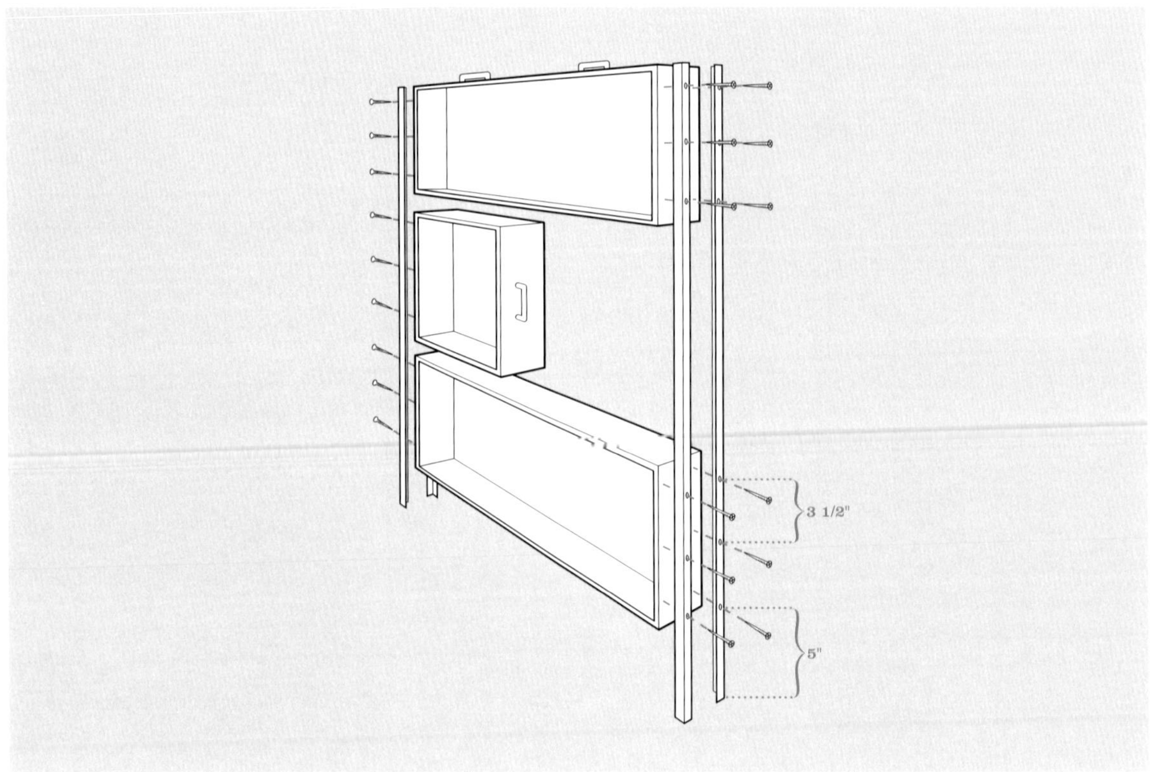

1 Settle on a pleasing arrangement of your drawers.

2 Measure the height of your configuration.

3 Measure and mark off one aluminum piece, 5" plus the height of your configuration (the extra inches will be the raised footing of the piece).

4 Use the hacksaw to cut the first aluminum piece. Use this piece as your template for making the other three.

5 Lightly sand the sharp, freshly cut ends of your aluminum pieces to avoid future injury.

6 Starting from the top of your configuration, position one of the aluminum pieces on the front edge of the top drawer unit. Make sure the top of the aluminum piece is flush with the top edge of the drawer unit. Hold in place.

7 Figure out an appropriate pattern for the screws that will connect the aluminum piece to the drawer and mark. We positioned our screws 3 $^1/_2$" apart and $^7/_8$" from the outer edge.

8 Using the $^1/_8$" drill bit, drill a pilot hole through the aluminum for each marked hole.

9 Using the $^1/_4$" drill bit, slightly drill the larger bit into the same hole so that the head of the screw will sink down into the wood when drilled.

The woods were lovely, dark, and deep, but he had promises to keep and miles to go before sleep, miles to go before sleep.

REDUCE YOUR TREE DEPENDENCE

Amazing Savings!

REUSE LUMBER

Demolition of homes and businesses results in enormous amounts of perfectly good lumber, the best of which is bought up and resold by salvage companies. Older wood is often of a better grade than what's commercially available today, and a set of creaking, pock-marked oak floorboards tells its own story.

RATION YOUR RAYON

The process of making this wood-based fabric (the less common acetate, triacetate, and Tencel are all made from trees as well) is toxic and wasteful, utilizing a slurry of water and chemicals to extract the necessary fibers from trees. In the end, only one third of the pulp is used to make rayon thread.

DUCHAMP'S CORNER

15 Alternate Uses for Chopsticks

Clock hands
Bird stand
Ear-wax remover
Hair bun holder
Furniture shim
Baked cake tester
Kindling
Drink stirrer
Candle-making wick holder
Mini drum sticks
Poo scraper
Tinker Toy stick stand-in
Eye poker
Puppet stand
Popsicle stick

10 Use the Phillips-head bit to tighten the screws into each hole.

11 Mirror your screw positioning for the aluminum piece on the backside of the drawer you are working on.

12 Repeat steps 7 through 11 for the next two drawers.

13 When the two aluminum pieces are screwed into the entire length of a side, flip the shelves to the other side and repeat steps 7 through 11. Make sure that the excess aluminum pieces (which will essentially be the feet for your shelves) end up the same length.

14 You may need to drive a couple of nails into the bottoms of the two top shelves to ensure there is no gap between the drawers.

15 Customize your shelves with paint if you wish. We did this to cover some of the wear and tear of the old shelves.

TIP: It's best to use drawers that have similar depths. We used two drawers from one dresser, and one from another unit.

DOOR MIRROR

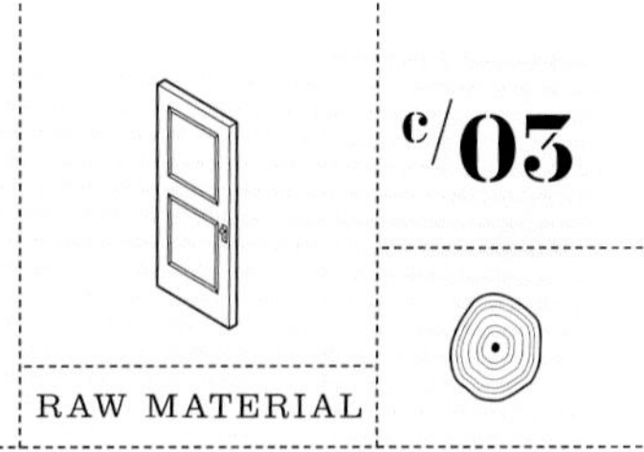

Sure, we'll fork over our paychecks for something that makes us feel good about ourselves. But for a household staple that reflects our image exactly as we are? Mirror vendors should be paying us! We take it as an even greater insult that the decorative wall mirrors on the market today are either of flimsy "modern" effects or fussy with baroque trim and abundantly overpriced. Using an old, salvaged door frame and a precut oblong mirror, it's easy to make a handsome, well-weathered looking glass for less than it'll cost you to improve your image at the hairdresser.

"They make alcohol out of wood in Russia," he said, in Russian. This was meant to explain why he always had a toothpick poking out of his mouth—right up until he turned out the light. Sixteen years he'd been here, and still the "mother tongue" was wagging over a little sliver of wood.

DOOR MIRROR

$210

INGREDIENTS

+ Old, paneled, solid-core wood door (ours had one long panel)
+ Mirrored glass, cut to size
+ 1/2" by 3' by 3' plywood
+ Nails
+ 1" screws
+ 2 keyhole screws
+ Heavy picture-hanging hardware (optional)
+ Picture-hanging wire (optional)

TOOLS

+ Tape measure
+ Handsaw
+ Sandpaper or router
+ Hammer
+ Hand drill
+ Hand circular saw
+ Wire cutters (optional)
+ Stud finder (optional)

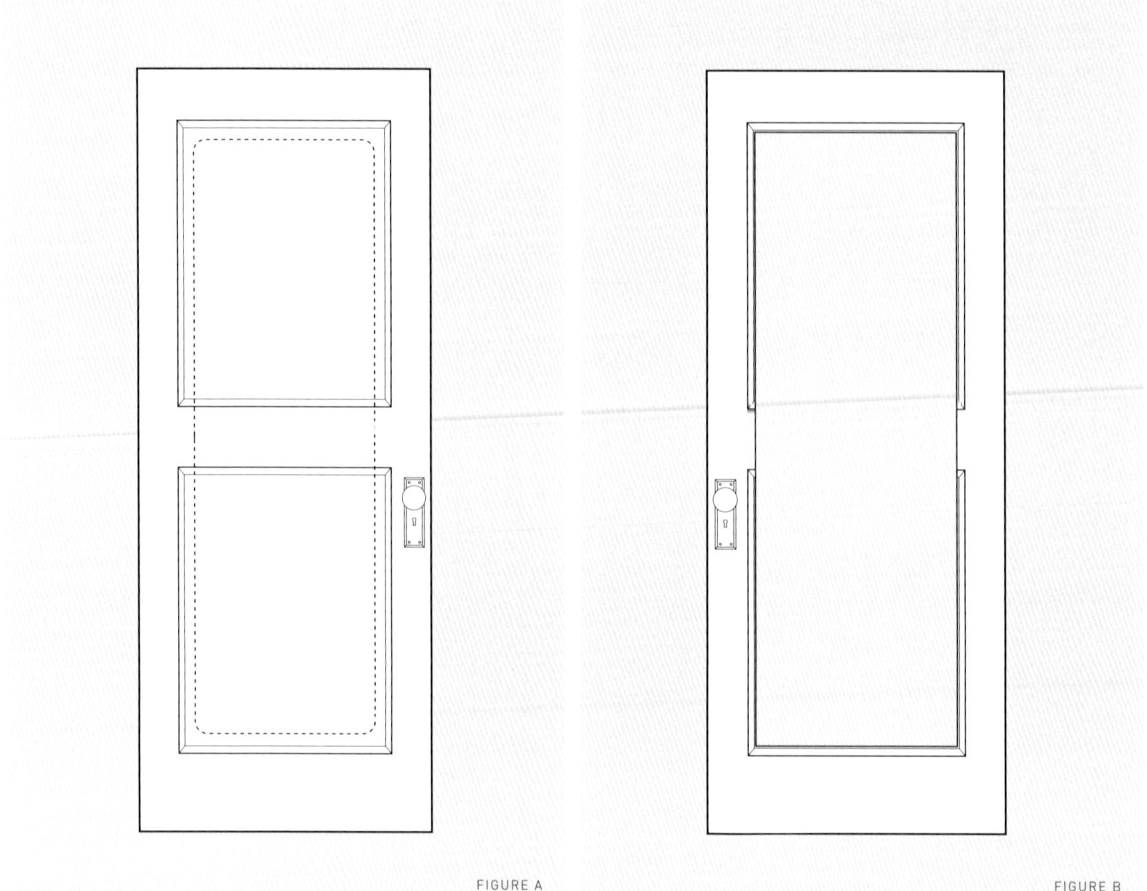
FIGURE A

FIGURE B

1 Measure and cut an opening 9" in from the outer perimeter of the door, or follow the panel dimensions and cut the opening 1" smaller than the panel size (Figure A).

2 Using sandpaper or a router, smooth out the raw edges of your opening.

3 Acquire a piece of mirrored glass that is 3/4" larger than the opening. (Having a glass retailer cut a piece to size doesn't cost much.) We recommend putting safety film on the backside of your mirror.

4 Carefully set your mirror in the backside of the door. It should nest comfortably in the existing panel insets (Figure B).

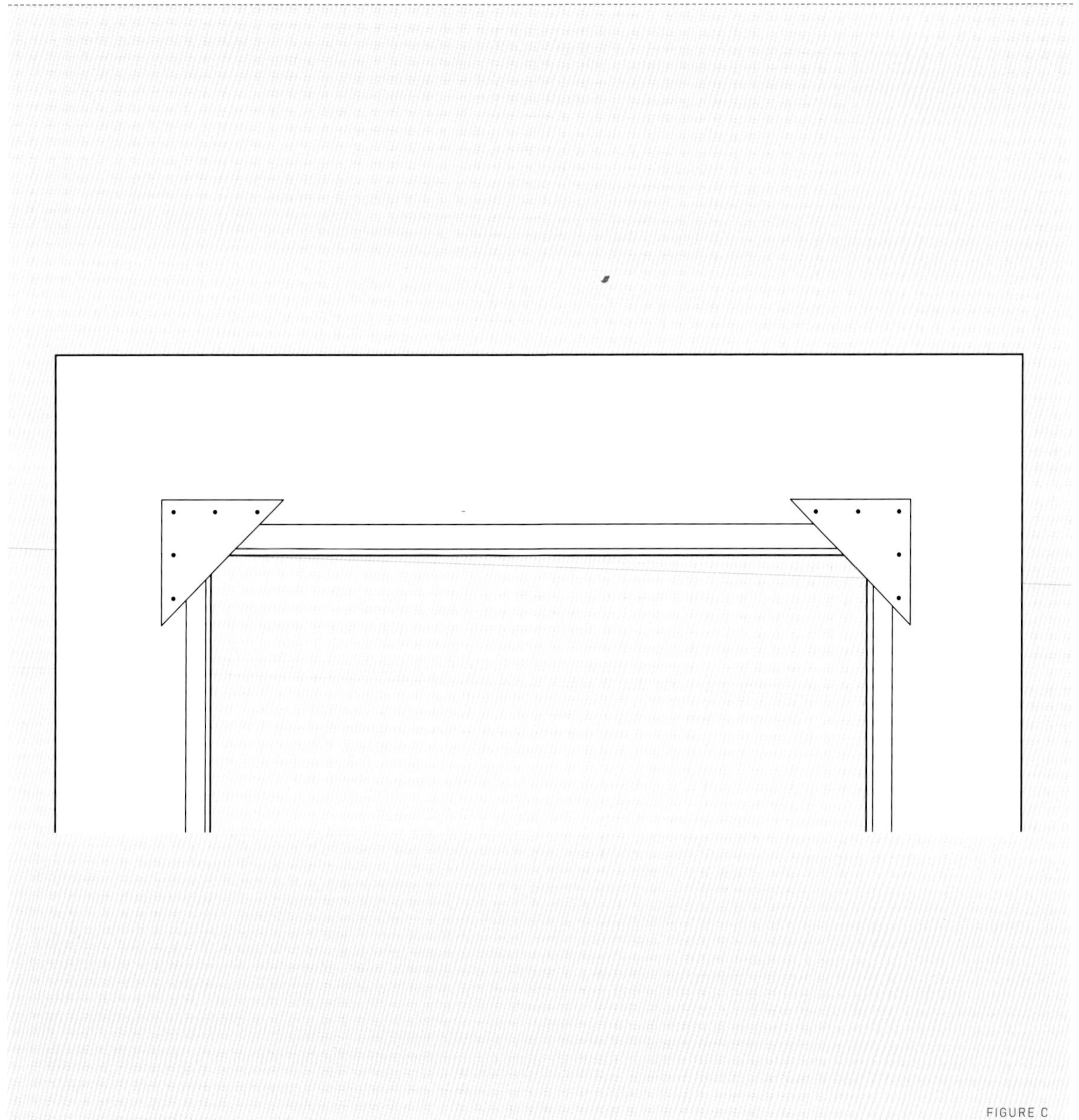

FIGURE C

5 Secure the mirror by cutting and nailing a triangle-shaped corner piece on top of each mirror corner. You may need to slip in a thin piece of soft shim if your mirror jiggles in the frame. (Figure C).

6 Attach keyhole screws to both the left and right sides of the back (we recommend a minimum of $1\frac{1}{2}$' down from the top).

7 Attach the appropriate amount of picture-hanging wire or lean the mirror against a wall. If you plan to hang it, make sure the screws are drilled into wall studs. A mirrored door is heavy—not the kind of thing you want toppling on someone's head.

VENEER LAMPSHADE

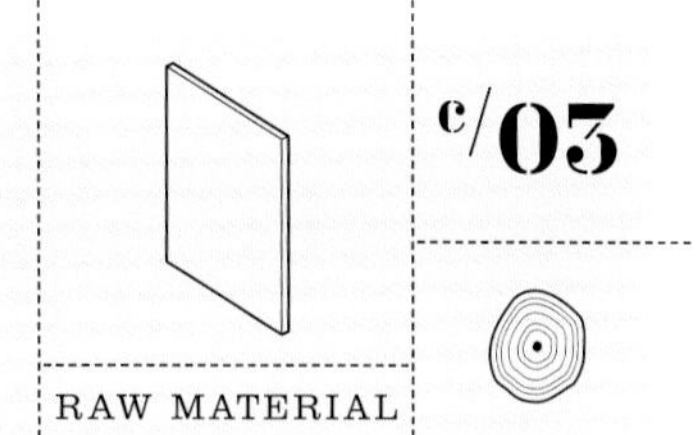

Everyone has an ugly lamp lurking around the house that's due for a face-lift (even last year's Ikea shades look dated to us now). Give yours a mid-century modern twist by stripping off the paper or fabric and replacing it with thin strips of wood veneer. Typically used for finishing plywood edges, veneer is incredibly easy to work with and gives off a warm, woodsy glow when lit. This is just a suggestion to get you going. We'll leave it to you to come up with other bright ideas: weaving veneer strips together, layering like shade shingles . . .

Millions of Christmas trees produced each year for American households: 34–36

VANEER LAMPSHADE

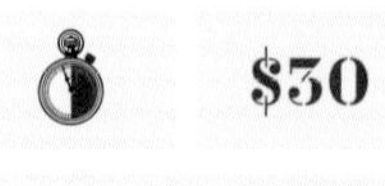

$30

INGREDIENTS

+ Old lamp with an ugly square-shaped shade
+ Wood veneer (size depends on your lamp shade)
+ 150 metal O-rings

TOOLS

+ White pencil
+ Utility knife
+ Scissors
+ Small hole punch
+ Pliers

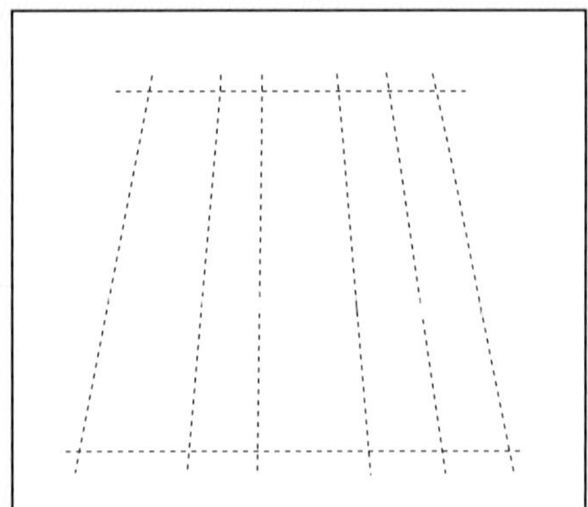

FIGURE A

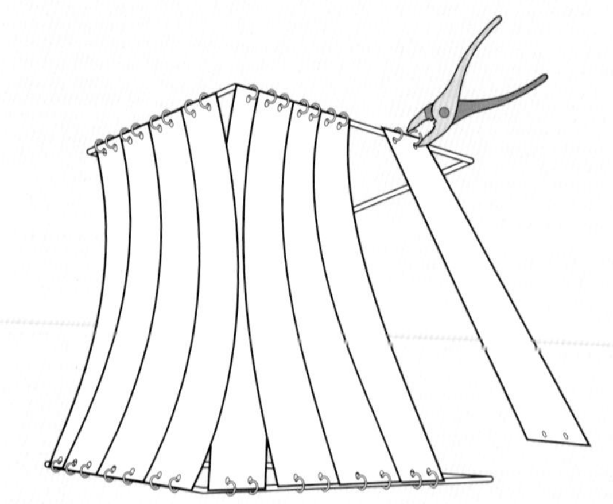

FIGURE B

1 With your white pencil, trace the shape of one side of the existing lamp-shade on the backside of the veneer. This will be your template for the other three sides.

2 Peel the ugly shade off its frame.

3 Cut out the veneer template with your utility knife.

4 Trace and cut out three more pieces, identical to your template.

5 Cut one piece into vertical strips of varying widths (Figure A).

6 Punch a couple of holes in the tops of your strips.

7 Use pliers to attach the strips to the top rim of your lampshade with the O-rings (Figure B). If your shade has a bottom rim, punch holes at the bottom of the strips and attach with the O-rings there too.

8 Repeat steps 5 through 7 for the other three sides of your shade.

9 Plug in, turn on, and enjoy the log cabin glow.

Years it takes an average Christmas tree to mature: 7–10

this is not a project

c/03 | WOOD

IF ONLY NOAH HAD LEFT INSTRUCTIONS

WE WOULDN'T BE SO WORRIED ABOUT THE GLACIERS

BUILDING AN ARK

WHY WE CAN'T SHOW YOU HOW

WE REALLY WANTED to teach you how to build a ship large enough to accommodate two of each of the world's creatures. It's been raining for days, and we've just witnessed the catastrophic tsunamis of late 2004. The glaciers keep dumping their extra winter pounds into the slushy arctic waters. All signs point to the inexorable: that a second flood will soon be upon us. It is with deepest regret that we admit to not having acquired a set of sufficient blueprints. You should know that we are committed to schooling ourselves in the marine arts this year. The seas will be high and swollen. We shall one day help you navigate.

Until then, take heart! We *have* included instructions for constructing a miniature version of Noah's venerable craft. This 27-inch floater can be used as a teaching aid, to accompany a reading of Genesis and our impending doom, or as a bathtub toy for grown-ups, good for hours of entertainment as you soak in bubbles scented with lavender and myrrh, enjoy a skein of wine, and nosh on matzo. Still, knowing you must be disappointed that this book does not include Noah's plans, and that, as dark times encroach, you will not yourself know how to build the original Save-Yourself! Ark™—the first woodworking project on record—we felt you deserved a lengthy, discursive explanation.

For centuries, intrepid explorers have made the pilgrimage to Mount Ararat, the peak in eastern Turkey where Noah's ark is said to have run ashore as the floodwaters receded. ("The ark rested . . . upon the mountains of Ararat," Genesis 8:4). They have gone with the hope that the most famous boat of all time, entombed in a Turkish glacier (at nearly 17,000 feet, the mountain wears a permanent ice cap—ideal for ark preservation) would have remained intact and could be dug up, spit-shined, and put on view at a museum or auctioned off piecemeal on eBay under the heading "Own a chunk of petrified Ark—salvation in a box!"

There were several accounts of vaguely craftlike remains found at Ararat—some made by pilots and treasure hunters of questionable character. *Life* magazine even ran a picture of a large, boat-like escarpment with the headline "NOAH'S ARK?" Geologists never found any evidence that an ark existed. But, more to the point, the ship we would have had you build to escape a watery grave suffered from serious structural problems, even in the absence of two of every beast and winged creature on Earth.

The Ark, as described by the Bible, is the largest pre-nineteenth century seagoing craft. It wasn't until 1884 that the Italian liner *Eturia* outdid its proportions,

NOAH'S MODEL WAS THREE STORIES HIGH, ITS DECK FOOTPRINT EQUIVALENT TO TWENTY COLLEGE BASKETBALL COURTS OR THIRTY-SIX TENNIS COURTS.

which are close to the size of the bulging steel ocean liners that carry groups of seniors to Alaskan destinations today. Noah's model was three stories high, its deck footprint equivalent to twenty college basketball courts or thirty-six tennis courts. The pertinent question here, in this chapter on wood, is how much lumber would such a feat of water-borne architecture require? If the average tree is 30 feet tall and 3 feet thick, just the outside of the boat would use up two hundred tree trunks. The planks would have to be cut thick, as the Ark would be a veritable zoo of animals, including some of the largest among God's creations, the elephants (critical crew members, as

they would remember the pre-flood world better than anyone else aboard). We can safely assume that it would take Noah and his sons the better part of a day to fell a tree, shear it of its branches, and haul it back to the Ark. For the amount of wood required, that would add up to nearly a year just to gather materials.

Did Noah have Norm Abram–like skills? Was he granted a short-term loan on omnipotence to complete the task? Sure, his sons Shem, Japeth, and Ham helped out, but without a Home Depot in Arabia, jigsaws were hard to come by. Whittling those trunks down into arcing bows and sterns would have demanded an awful lot of patience. The largest ship built out of wood, the U.S.S. *Wyoming*, took years to complete, required modern machinery, and employed thousands of salts sweating it out in a dockyard. The *Wyoming* was shorter than the ark by more than 100 feet.

Of course, they lived a lot longer in biblical days (Noah was five hundred years old when he started the project), so they had more time to kill. But you can see how we didn't want to saddle you with a project that would eat up most of your weekends into the foreseeable future and still remain unfinished when you breathe your last breath. Instead, think about ways you can help prevent the next flood (be good unto others and unto the ozone layer), practice some shipbuilding skills on the miniature ark we've included here, and hope for the best.

THE ORIGINAL, IF SLIGHTLY SKETCHY, INSTRUCTIONS

Genesis 6:7 *And the LORD said, I will destroy man whom I have created from the face of the earth; both man, and beast, and the creeping thing, and the fowls of the air; for it repenteth me that I have made them.* **6:17** *And, behold, I, even I, do bring a flood of waters upon the earth, to destroy all flesh, wherein is the breath of life, from under heaven; and every thing that is in the earth shall die. But Noah walked with God, and he was to be spared.* **6:18** *But with thee will I establish my covenant; and thou shalt come into the ark, thou, and thy sons, and thy wife, and thy sons' wives with thee.* **6:20** *Of fowls after their kind, and of cattle after their kind, of every creeping thing of the earth after his kind, two of every sort shall come unto thee, to keep them alive.*

ARK GAMES "FOR PIOUS YOUNGSTERS"

→**The Seventh-Inning Stretch of Creation:** The children read Genesis 6:7, in which God announces he will abolish all creatures roaming the Earth and start over. Likewise, they stop their baseball game, break for snacks, switch teammates, and start over.

→**Pickle Night:** Children learn about Noah's creative food preservation techniques and guess which item in their lunch bags will last the longest.

→**Noah's Ark Match-up:** Spin the boat (for teenagers only).

c/03

Su Mo Tu We Th Fr Sa

Why Don't I Do This Every Day?

TESTIMONIAL # 0036

NAME: Chelsa Robinson

FROM: Calgary, Alberta

OCCUPATION: ReadyMade Product Manager

Chelsa, why don't you do these things every day?

I'm not one for too much hoopla, so the twig candle thing has a certain no-fuss charm. Just don't leave it unattended! I suppose you'd need a dining room table for one of those anyway. Or a sideboard—yeah, right! As for the chopstick clock, it has a nice simplicity, but I'd kick it up a notch. (Hello, casino dice?!) But let's not be too hard on ourselves. If a clever, inventive, and environmentally responsible reuse project has to wait until tomorrow, it can—you crazy overachievers!

1

TWIG CANDELABRA

INGREDIENTS

+ Inspiringly shaped branch with twigs
+ 1/2" finishing nails
+ Short and slender candles

TOOLS

+ Sandpaper
+ Hacksaw
+ Hammer

01 Collect a cool twisted branch.

02 Remove the bark with sandpaper.

03 Saw off the end at an angle that allows for the branch to sit stably in an upward-reaching manner.

04 Carefully hammer two finishing nails where you plan to secure candles from the bottom up so the points of the nails are sticking up.

05 Balance your branch and candles as you select twig locations. Be mindful of keeping candles apart from each other and away from low-arching parts of the branch.

06 Insert candles and have a romantic dinner.

Why don't we?

2

SAWHORSE TABLE

INGREDIENTS

- One pair of sawhorses (we found a great folding version with a horse illustration on it online at Home Depot)
- Large piece of 1" thick plywood, an old undivided door, or a custom-cut piece of glass
- Small pieces of thin carpet padding (optional)

01 Arrange the two sawhorses apart to hold your table surface.

02 Set your tabletop surface of choice on top. (You may want to place a couple 1" by 2" slivers of thin carpet padding in between to prevent your surface from slipping.)

We really should, you know.

3

CHOPSTICK CLOCK

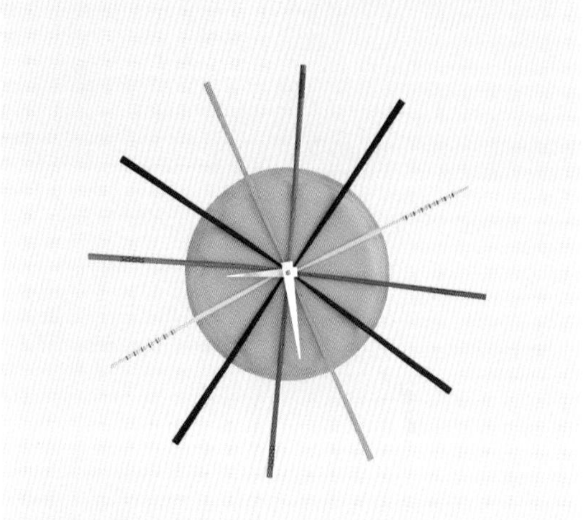

INGREDIENTS

- 6 pairs of chopsticks
- Clock kit (available at *www.readymademag.com*)
- Paper plate

TOOLS

- Hot glue gun
- X-Acto knife

01 Use the glue gun to affix the chopsticks to all twelve time positions around the backside of the paper plate.

02 Use the knife to make a 5/16" hole in the center of the plate.

03 Assemble the clock mechanism and mount through the center of the paper plate.

04 Hang your clock on the wall. Tell time.

Tarry no longer, chop chop!

Chapter: **04**

METAL

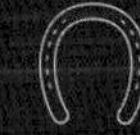

SLAG-HAPPY

Bronze Age, Iron Age, Industrial Age: History is all about how good we've become at working our ore. Without blacksmiths we'd still be living like the Flintstones, braking our cars with our bare feet and carrying clubs to clobber foes. Metalsmiths were important everywhere, and from the very beginning. How do we know? Because they show up in the cross-cultural pantheon of gods. The first ironworker is given ink in the Old Testament: "Tubal Cain, an instructor of every artificer in brass and iron" (Genesis, 4:22). The Greek god Hephaestus, born lame, turned out to be a fantastic forger. In the African Sudan, Nzeanzo is the smithing god. In ancient Persia, the god of the Vrahran Fire punishes the evil done by man and demon by foisting upon them a sacred combination of sixteen flames, most of which belong to those in the metalworking trades. And British rock gods The Smiths rose to fame in the late twentieth century for creating supremely weepy, high-minded songs about vegetarianism and love gone wrong.

But anything that inspires worship can be dangerous, too. As the Romans set out to conquer the world, they were being poisoned at home. It's thought that lead may have been to blame for the empire's decline, since it was used to deliver water and indoor plumbing to those who could afford it. In one of the greatest reversals of fortune of all time, Rome's patricians died off early and often (the lead from the pipes leached into the water supply), while plebs who drank straight from the well lived long, hydrated lives.

A few dead royalty did nothing to diminish the demand for metal in every walk of life, however. During the Middle Ages, being a smith was like being a lawyer or a shrink—everybody needed one. The Church wanted custom spires and confessional grilles and chancel screens and engraved doorplates that made a sanctimonious *thud* as they closed people in for prayer. Then there was the matter of shields and battle-axes and horse bridles and chain mail to protect the knights of the Crusades. (*Mail* means hammered, by the way, not delivery to foreign lands. We know you wonder about these things.) Then there were the domestic goodies: iron or lead goblets, roasting pans, fire pokers, and the common nail. Do not forget that the nail is a thing of rare beauty.

The rare, shiny metals that occur in crude form in nature include gold, silver, and various members of the platinum family (ruthenium, rhodium, palladium, osmium, iridium). They were welded entirely by hand and shaped with hammer and anvil. Steel, a mixture of iron and carbon, wasn't even in the picture yet outside of China (where everything gets invented centuries earlier). Refined gold or silver in bulk (ingots) is called *bullion*. We can't help but wonder why we call those tiny auric cubes of condensed chicken or vegetables that make soup stock by nearly the same name. As with all things, the value of bullion vs. bouillon is relative; you can't fill your belly with a hunk of gold, and soup cubes won't buy you much in the open market.

Being a smith was like being a lawyer or a shrink—everybody needed one.

THE GOLDEN AGE OF METAL

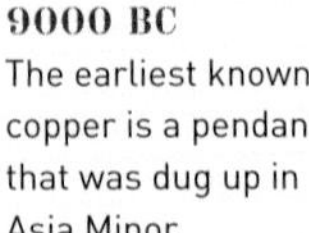

9000 BC
The earliest known copper is a pendant that was dug up in Asia Minor.

4000
Thought to turn black when exposed to poison (hence its use to make chopsticks for the paranoid nobles of China), silver is soft like its cousin gold and mostly turned into jewelry and filigree.

3500
The fist of the divas, early Egyptians use lead galena (lead sulfide) as eyeliner. Lead also, they found, made excellent piping for the transportation of water.

900
The Chinese learn that by pumping oxygen into the blast furnace, they can remove carbon from iron, creating the first steel. They call the process "hundred refinings method," as it is repeated one hundred times, and prize the output, shaping it into keen-edged swords.

AD 1340
The Chinese start setting off fireworks with gunpowder in them to celebrate their new year. The bigger bang of the bronze cannon is invented in Europe.

1447
Presses in Germany begin to use engraved copper plates for printing.

1855
Brit Henry Bessemer develops the first process for mass-producing steel. The steel manufacturing technology of today is a refinement of his inexpensive method.

1876
Voice transmission occurs along copper telephone cables.

1879
Electric train motor is developed in Germany. The Teutons boast all the early motor skills.

1886
After a century of independence and tired of the fish-and-vinegar smell still present in their vestments from colonial days, Yanks develop a washing machine with an electric motor.

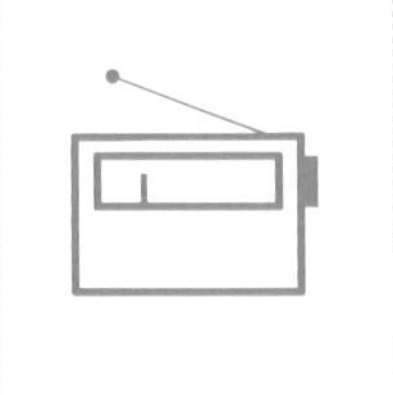

1906
Electric radio with copper and brass wiring and components is invented.

How copper pendants in Asia Minor

ended up shrinking down

to copper wires on microchips.

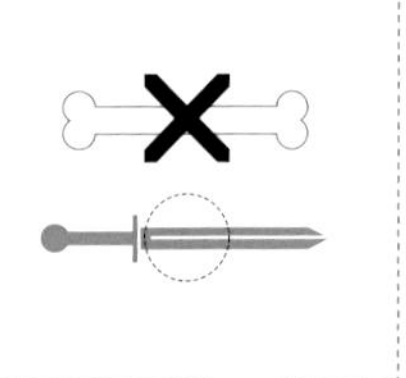

6000
Ancient societies found plenty of copper and gold as they carved their stone cities out of mountains. How could they miss the dazzlingly shiny stuff as it glinted out from cave walls like bad hotel wallpaper? Back in the Stone Age, the jewelry industry was born when the first of the Forty-second St. sales guys discovered that gold could be hammered into sheets and wires.

1500
Small amounts of smelted iron were found in meteors in ancient times and wrought by crafty Far Easterners into bells and gongs—among our first musical instruments. Bones, though good for clobbering, felt so Stone Age once iron came on the scene. With its deflective shielding properties that kept people alive longer, iron made warfare more fun.

700
The advent of pocket change: the use of bronze as a means of payment becomes widespread in Asia.

1510
Brass is used for the cases and gears of pocket watches, and later for pendulum clocks and gears.

1608
A telescope made from brass spies the heavens over Holland.

1821
The Brass harmonica and, in 1841, the brass saxophone continue the tradition of musical copper.

1837
Electric telegraph transmissions are made along copper cables in the United States. In 1851, the first underwater telegraph cable was laid from Dover to Calais.

1881
First electric lightbulb fed by copper wires. Later, alternating current along the wires was developed as an efficient means of transmission.

1885
German engineers, who will become famous for their automobile prowess, make motorcar components out of brass.

1917
The first electric drills made from a motor fitted with copper windings.

1954
Solar cells, using copper wiring, convert sunlight to electricity.

1998
Copper circuits on microchips increase processing speeds.

2005
Wireless transmissions, now widely in use, begin to reduce our age-old reliance on copper.

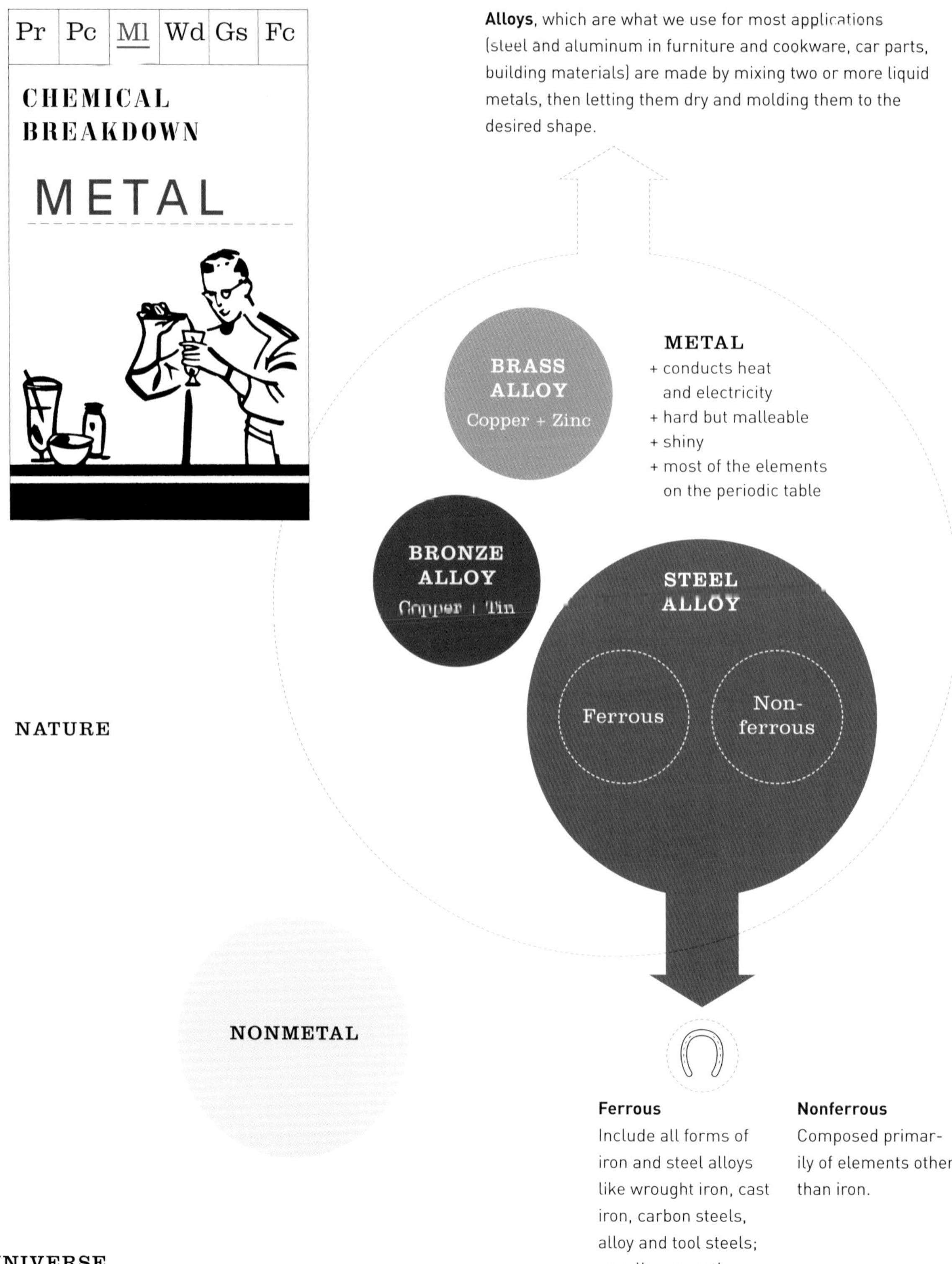

Alloys, which are what we use for most applications (steel and aluminum in furniture and cookware, car parts, building materials) are made by mixing two or more liquid metals, then letting them dry and molding them to the desired shape.

Ferrous
Include all forms of iron and steel alloys like wrought iron, cast iron, carbon steels, alloy and tool steels; usually magnetic.

Nonferrous
Composed primarily of elements other than iron.

FROM ALLOY TO ZINC: A GLOSSARY OF TERMS

Alkali metals: In comparison to the other metals, they are soft, and have low melting points and densities. Tarnish rapidly (even in dry air). Lithium, sodium, potassium, rubidium, cesium, francium.

Alkaline-Earth metals: Generally softer than other metals, react readily with water, powerful reducing agents—alkali metals exceed all of these properties. Form divalent compounds. Beryllium, magnesium, calcium, strontium, barium, radium.

Brass: *(copper + zinc alloy)* Brass monkey was a metal plate on a warship upon which cannonballs were stacked for use in battle. In cold weather, the metal contracted, causing the balls to fall down. Also an early Beastie Boys song.

Bronze: *(copper + tin alloy)* Third place. Also, the first alloy and a major discovery, as bronze is harder and stronger than copper and was used to make tools and weapons (all you really needed back then, besides a pelt and a good rainy season).

Cesium: A silvery-white metallic element that is commonly used in the photoelectric cells that act as transmitters in cameras, televisions, and traffic lights.

Cobalt: A tough, lustrous, silvery-gray element that has the privilege of being harder than iron. It can be found in steel alloys, paints, and glass.

Copper: You've seen the piping and the cookware. Copper is ductile and corrosion-resistant, and runs a high temperature when submitted to heat. Twenty years ago, it was thought that modern society would run out of copper wire for transmitting telephone calls. Now that cell phones are taking over the world, we don't need the stuff as much.

Gold: A heavy, soft, ductile, malleable element, gold is found in quartz veins, in nugget form, and as a more fine dust panned out of riverbeds. Because a gram goes a long way (it can be stretched into a very long, thin wire), it's used most commonly for jewelry, and comes in white, yellow, or green hues. Because it's so soft it must be mixed with copper, silver, nickel or palladium—one carat means one part gold in twenty-four parts baser stuff.

Iron: The cheapest and most common, useful, and important metal. Used for making steel and many other alloys. We have become dependent upon it for our cars, steel beams, rails, dishes, and bridges. Our bodies need iron, too. It transports oxygen through the blood to the rest of your body. It can be gotten from fish, meats, whole grains like wheat, and beans like lentils.

Lead: By no means pencil thin, lead is the heaviest and softest of the common metals. Used as a filler, it resists corrosion, and is ductile, highly malleable, and glossy looking. It's found in the conductive heart of batteries, in the paint that kids shouldn't lick; and in explosives.

Mercury: Has the distinction of being the only metal that exists in liquid form at room temperature, which is why it moves up your thermometer. Used for its ability to dissolve silver and gold—the basis for plating technologies.

Nickel: A hard, malleable, ductile, silvery-white metal. Used in magnets, stainless steel, and armor plating. Our five-cent piece was named after it, being the coin's chief ingredient. Nickel produces green color in glass.

Platinum: An element on the periodic table that's heavy, malleable, ductile, precious, steel-colored, and resistant to corrosion. Platinum is used in jewelry, lab equipment, catalytic converters in cars, and dentistry. When heated, it can soak up then release hydrogen gas, so platinum may help us save ourselves from cooking the Earth in fuel-cell cars. Alas, we're still decades away from clean commutes.

Silver: A white metallic element, ductile, malleable, with the highest thermal and electric conductivity of any substance. Used for coinage, jewelry, electronic parts, and wire. Of total world supply, 72 percent is used for monetary purposes.

Steel: *(iron + carbon alloy)* The stainless variety is what we most often come in contact with (using it to feed ourselves), but this tough alloy is everywhere—it may be our most commonly used metal.

Tantalum: A rare, brittle, lustrous, heavy, gray metal. Used in alloys having high melting points, and in aircraft and missile parts. Alloys used for springs, gun barrels, saws, dental and surgical instruments. Used with electrical wires because of its tendency to follow electricity in one direction.

Zinc: A shiny, bluish-white metallic element. Used in many alloys, brass, cables, fuses, and die-casting. Also an important metal to ingest, and commonly used in medicines (zinc lozenges to sooth a cold, for example).

HUBCAP FOUNTAIN

RAW MATERIAL

c/04

Few things inspire nostalgia like the old Motor City insignia. But instead of rolling off the lot in a 1970s-style emissions-heavy clunker, try out this hubcap project. We found a large Cougar model and a smaller cap from a Chevrolet among stacks of hubcaps at our local salvage yard. In the past we've drilled them through the center and fitted them with clock parts, and even considered them as food bowls for Fido. But when this multitiered backyard sculpture with water driven through it came to mind, it fast became an obsession. No coolant required.

Number of years an average home's electricity would be powered by recycling one ton of aluminum: 10

HUBCAP FOUNTAIN

$105

INGREDIENTS

+ 2 hubcaps, one larger than the other (ours had diameters of 10 1/2" and 15")
+ 1/2" plumbing hex bushing
+ 2 1/2" plumbing couplings
+ 2 1/2" x 8" plumbing nipples
+ 2 2' heating ducts
+ P140 water-fountain pump
+ 12"-diameter plastic bucket
+ Heavy-duty garbage bag
+ 2' of 1/2" plastic tubing

TOOLS

+ Hand drill
+ 7/8" titanium drill bit
+ Metal shears

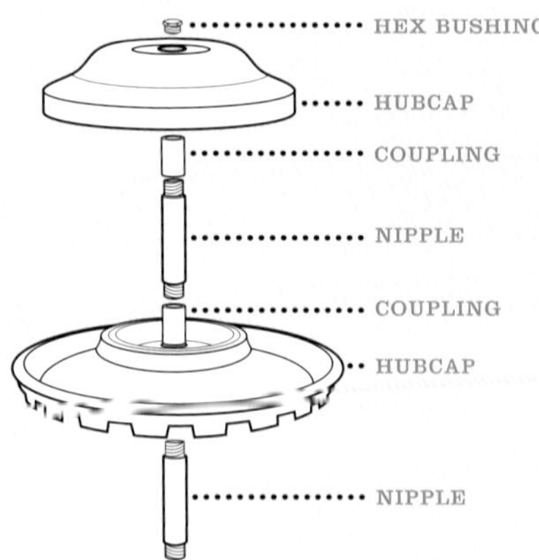

TOP HALF

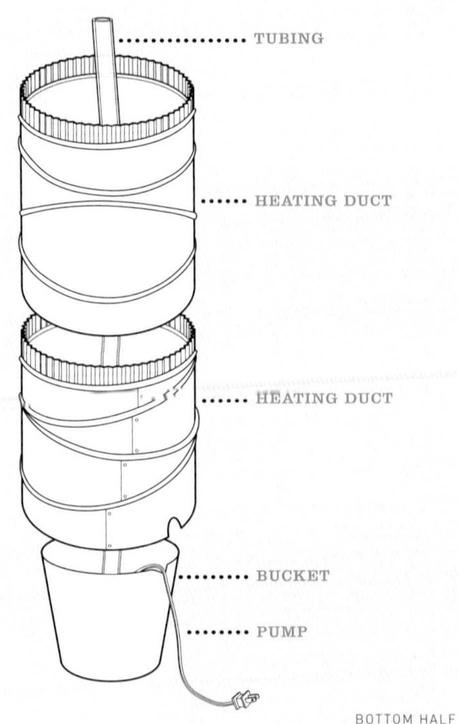

BOTTOM HALF

1 Drill a 7/8" hole in the center of each hubcap.

2 Insert the 1/2" hex bushing nipple into the top (from the top down) of the smaller hubcap so the top ridge rests snugly on top.

3 Hand-twist on one of the 1/2" couplings to the threads of the nipple on the underside of the top hubcap.

4 Hand-twist one of the plumbing nipples into the bottom of the coupling.

5 Hand-twist on one of the 1/2" plumbing couplings to the threads of the nipple you just put on.

6 Insert the other 1/2" by 8" plumbing nipple from the bottom up into the bottom of the larger hubcap.

7 Hand-twist the threads of the second fitting into the bottom of the nipple above. This hubcap should sit securely in place.

8 Stack your aluminum heating ducts on top of each other. They are self-fitting, so marry the female end to the male end. This will be your fountain stand.

9 Drill two holes, 1/4" from the bottom edge of your fountain stand.

10 Use your metal shears to cut the two holes you just drilled into an opening so the electrical plug to the fountain pump can fit through.

11 Fill the bucket half-full with

Number of aluminum cans that must be recycled to save enough electricity to operate a TV for three hours: 1

BICYCLE WHEEL FAN

Duchamp and Picasso saw it, and so do we—bike wheels, with their thin aluminum spheres and spokes, are as much a work of art as they are a mode of transportation. Paying homage to this much-loved nineteenth century invention—its ability to translate pedal power into the nearest thing we know to flight—we came up with another way to feel the wind in your hair. But when it came right down to it, the blowing-power was pretty feeble compared to your garden variety store-bought brand of fan.

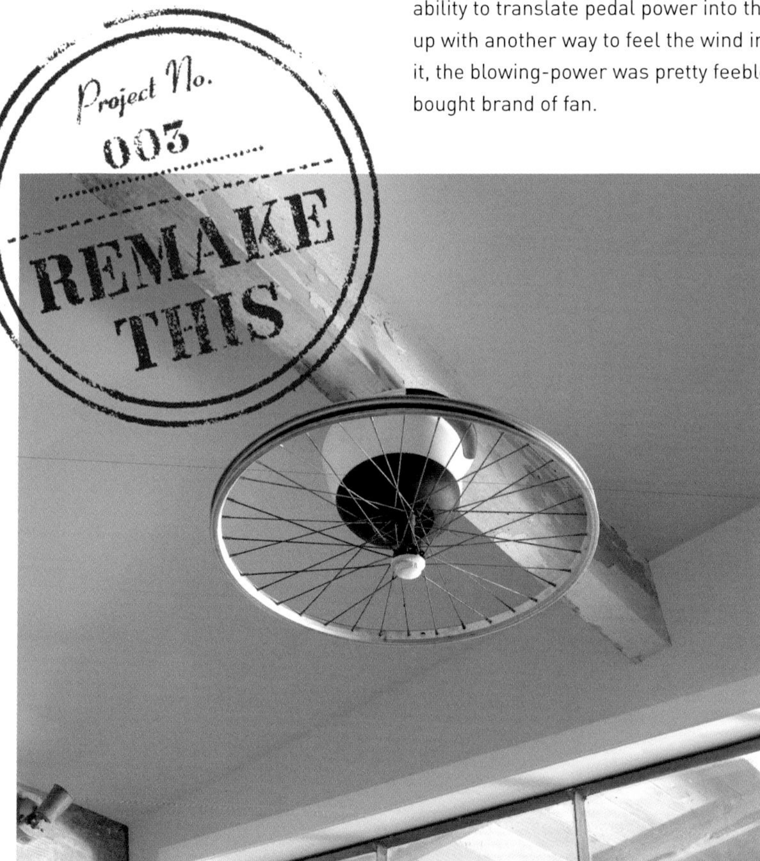

YES

1 Bike wheels, with their lattice of spokes, are as much a work of art as they are a mode of transportation.
2 You can customize your fan by hanging different materials from the spokes.

NO

1 Not all wheels fit the salvaged fan wheel you'd be using.
2 It was a challenge to conceal the motor in a good-looking way. We came up with a plastic bowl. Yikes.

water. Place the bucket in the fountain stand, set the fountain pump in the bucket, and pull the plug outside through the opening.

12 Make a soft drain out of the garbage bag that connects from the hole in the hubcap and drains to the bucket. The bladder is used to channel water into the bucket.

13 Connect one end of your 2' plastic tubing to your pump.

14 Hand-feed the other end of your plastic tubing up the bottom of the last fitting, under the larger hubcap.

15 Set the fountain structure on top of the fountain stand. (Our hubcap fit almost perfectly on top of the stand, but you may have to make some adjustments based on the size of your found materials.)

16 Plug in, watch it gurgle, toss in a few pennies.

NOTE: Our larger (bottom) hubcap has a concave perimeter that can hold water. You may have to modify the design if yours can't act as the fountain's basin.

I'M
HOME

COAT-HANGER WINE RACK

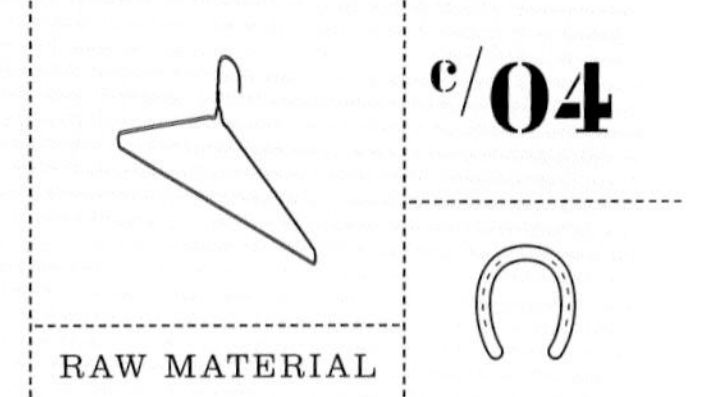

Fed up with coworker complaints about too few coat hooks in the factory, an employee of the Timberlake Wire and Novelty Company in Jackson, Michigan, bent some wire into two ovals and twisted the ends together to form a hook, thereby giving the world the coat hanger. That was in 1903. Today, the boomerang-shaped hangers are the pennies of the clean-clothes industry—so valueless, they're given away for free. What to do with the stack you've been accumulating? You can return them to your cleaner or, for you oenophiles, try bending the wires into this simple wine-storage solution.

Average number of six-packs of aluminum cans recycled by American families per year: **150**

COAT-HANGER WINE RACK

$15

INGREDIENTS

+ 36" wood closet rod, or 1 3/8" wood dowel
+ Wood closet-rod sockets
+ 8 wire hangers
+ 2 pieces of plywood, 3/4" by 12" by 13"
+ 2 1 1/2" wood screws

TOOLS

+ Pliers
+ Hand drill
+ Hacksaw
+ Ruler

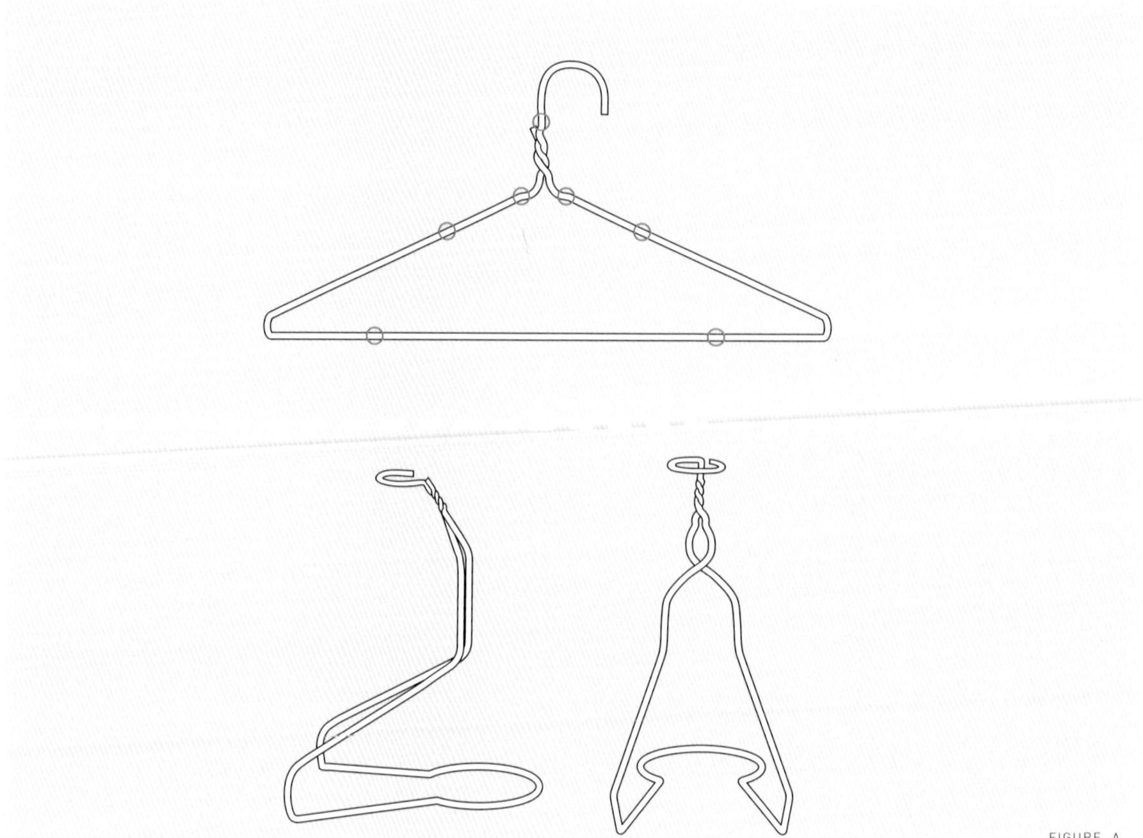

FIGURE A

1 Use your hands and the pliers to bend your wire hangers (Figure A).

2 Mark a spot in the top center of your plywood for mounting the closet-rod sockets.

3 Drill wood screws in from the outside of the plywood, through the sockets and into the rod. Drilling guide holes first will make the job easier.

4 Notch the rod with a hacksaw to keep the hangers in place. The part of the hanger that sits on the rod should be 1" wide. Each set of notches should be 3" apart. A 36" rod will accommodate eight sets of notches (Figure B).

5 Fit the hangers into the notches on the rod. The weight of your wine bottles will hold them straight.

6 Find a location that's not in direct sunlight to place your rack, and slide in your bottles of liquid gold.

How many days it takes Americans to use enough steel and tin cans to construct a pipe running from Los Angeles to New York and back again: 1

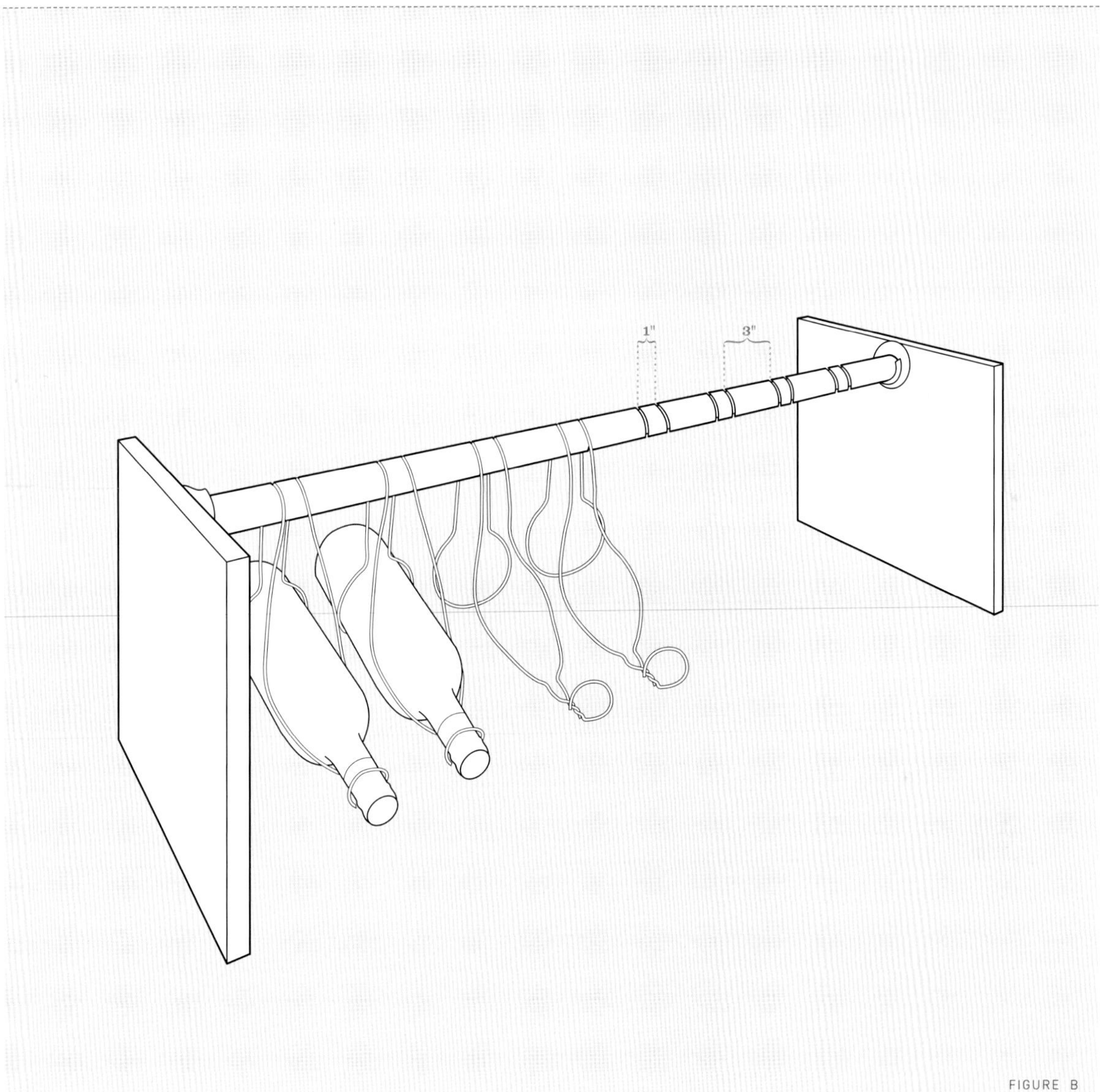

FIGURE B

NOTE: Though this version of the wine rack is designed to be freestanding, you can install it without the plywood sides anywhere a closet rod can be hung. Take advantage of the seldom-used storage space beneath your hemlines!

this is not a project

∩ | c/04 | METAL

FEAR HAS A THOUSAND FACES, FROM THE TRAPPED CORPORATE LACKEY TO THE SMILING CLOWN

HOW TO
STEEL
YOURSELF AGAINST
CERTAIN FEARS

FIRST: Don't knock fear. Fear can be a great motivator. Homo sapiens would be a wild bunch, driven to frothy-mouthed pursuit by gluttonous appetites, were it not for the fear of a punishing God. And it could be said that all of human achievement—great works of art, literature, scientific discovery—have been spurred by mortal terror, whether that be just an abhorrence at the thought of never enjoying a sandwich again or, for believers, the threat of eternal hellfire. Just the suggestion of death (a sudden heart flutter, mid-flight turbulence) goes a long way toward making people behave. And even minor frights have their perks. Career anxiety, for example. It's not altogether bad that the fear of dying a slow death by PowerPoint would drive someone to write their own pink slip and risk doing something completely different.

There's nothing more hopeful than someone who decides, suddenly, to change everything. To ignore the alarm clock, call in sick, go to the zoo. To then crouch down in front of the chimp cage and watch the female groom the male in a tight arm clasp, knuckle-walk to a shady spot, and then sit back on her bum cradling her fattened, middle-aged belly. Not with the languor that comes from living free in the wild, but from the exhaustion that comes with gadding about all day, waiting to be fed. An emotional palette of gray where red and green desires once were. Our man at the zoo watches the chimps and says to himself, *My God, am I no better than a monkey in a cage? I must change everything!* Such catalyzing fear of captivity, boredom, and the slow, inexorable climb to the treehouse in the sky goes a long way toward stirring up the courage to set off on a new course.

Unless, of course, one grows to love his captors, who keep him safe from certain threats. Rogue warlords throughout history have used our fear of others to their own perfidious ends. Here's an abridged history of haters.

A TIMELINE OF FEAR AND LOATHING

3000–100 BC
The pharaohs of Egypt consign slaves to back-breaking construction and various other custodial duties around the palace (in the Old Testament, the story is told with sea-parting flourish in Exodus). How do they force their subjects into labor? By issuing heartless dictates, like making the Israelites toss their first-born males into the river. Sure it was cruel, but without such intimidation, we would have no pyramids!

AD 1000–1400
Staving off an advance from Muslim invaders, the Eastern and Western arms of the Church, led by emperor Alexius I (in Istanbul, not Constantinople; cf. They Might Be Giants) and Pope Urban II (in Rome), call for crusades that will kill all infidels in their path and attempt to retake Jerusalem.

1400
Niccolo Machiavelli green-lights centuries of tyranny with his popular guidebook for rulers, *The Prince*. During the next five hundred years, the slim volume's most famous axiom, that it's better to be feared than loved, is followed to the letter by innumerable heads of state.

1500s–1700s
American colonists kill off tens of millions of natives, stretching from the Canadian Arctic to South America, by means of death marches, introduction of disease, scalping, poisoning, and murderous land grabs.

1788
The invasion of Australia by Europeans cuts a swathe through the population of Aboriginals, which drops from around 750,000 to 30,000 by 1911. On the shores of Europe, Robespierre goes wild with the guillotine in a grisly reign of terror—a futile attempt to put down the revolution—giving rise to dwarf-like dictator Napoleon Bonaparte, who sends hundreds of thousands of musket-slung grunts out to conquer the world for France. Bonaparte, in the Italian tradition, decrees the only two levers for moving men "self-interest and fear."

Late 1800s
Chinese slaughter at the hands of the Japanese; Ottoman Turks' murderous holiday in Armenia.

1900s
The twentieth century's catalog of atrocity may be the grisliest of them all: Stalin's gulags, Hitler's genocidal robots, America's trigger-happy bombing (one of them nuclear) of various nations, Pol Pot and the Khmer Rouge's slaughter in Cambodia, Baby Doc Duvalier in Haiti, a Yugoslavian battle over land and faith, the butchery of 500,000 Tutsis in Rwanda, and a tribal battle over land in Sudan/Darfur . . .

AFTER that brief throat-clearing, demonstrating how fright can motivate both good (ethical conduct, great works of art, the drive for change) and evil (annihilation of the other), we can move on to the instructive portion of our program. While you should never seek to eliminate the productive kind of anxiety—that which stirs you to do good in the world—we'd like to suggest a few tips for how to steel yourself against fears that lead to ruin. Follow this plan and set yourself free:

Xenophobia: "Fear of the strange," as the Greeks would have it, brings out the worst in us. In the not-us we see the threat of change, of being bettered, of erasure. But here's a thought: To them you are equally scary. No matter what your color, creed, or preference in cars, you are terrifying in your status as other, your you-ness. So, how to conquer the dread of strangers? Follow the Buddhists, who are among the nicest people we know. These steps will help you to attain Bodhicitta, or Buddhahood—a deep compassion for the suffering of other beings.

MIND TRAINING TO GENERATE BODHICITTA

Note that these exercises will not cure you of your fear of strangers overnight. It's advisable to meditate on this several times if you're really afraid.

Visualize three people standing before you: on the left a good friend, in the middle a stranger, and on the right an enemy.

- Concentrate on the friend and examine your feelings toward him.
- Now concentrate on the stranger and examine your feelings toward him.
- Now concentrate on the enemy and examine your feelings toward him.
- Return to the stranger and realize that this person can easily become your friend or enemy in the future.
- Now look at the friend and realize that this person could become your enemy in the future by betraying or hurting you.
- Now look at the enemy and realize that this person may become your friend in the future by helping you.
- Again look at your friend and feel your strong love and appreciation.
- Now look at the stranger and direct that appreciation toward this person.
- Again look at your friend and feel your strong love.
- Now try to hold this feeling while looking at your enemy; is it really impossible to feel some compassion for the one you revile?
- All three—friend, stranger, enemy—are equal in trying to become happy and trying to avoid suffering.

(Adapted from *www.buddhism.kalachakranet.org*.)

AEROPHOBIA: Fear of Flight

Why be afraid of air travel? Well, here's one reason: You're 35,000 feet in the air, going 550 miles per hour, and strapped in to a 147,000-pound aluminum tube with complete strangers at the controls. Then again, it's hard to deny the appeal of soaring above billowy clouds, of watching the Earth recede into a quilt of farmlands snaked by rivers, with the sun glinting off swimming pools and tiny cars. Plus, except for time travel, it's really the only way to get anywhere fast. Here are a few facts to help combat your fear of flying:

→System failures: According to the FAA, there are 1.85 fatal crashes per million departures due to system failure, weather, or pilot error. In other words, your odds of dying in a plane crash, if you fly only once or twice a year, are roughly equivalent to your odds of getting hit on the head by a falling plane. Planes have lots of backup systems. If one component fails, another takes its place. If a 747 gets a flat or two, it's no big deal—the plane has eighteen tires. If the jet was built in the last ten years or so, it also has two or three autopilots that can take over all functions.

→Midair collisions: According to the Bureau of Transportation Statistics, in 2002 the number of pilot-reported midair near-collisions was 180. Air controllers give each plane a 10-mile-wide private highway in the sky. No other plane is allowed to enter that lane.

→Bumps: Boeing and the like build planes that are tested to withstand six to seven g's of force. That's like flying into a tornado. Turbulence, or "chop," is measured in terms of gravity. Point-four g's of force is considered severe, and federal regulations require planes to be able to fly through at least two g's without trouble. Even better: Standard practice is to avoid all thunderstorms by at least 20 nautical miles.

→**Unusual sounds and sensations:** According to the National Business Aviation Association, at 1,000 feet the plane should accelerate to final segment speed and retract flaps, then maintain quiet climb power until reaching 3,000 feet, at which point the pilot can resume the climb. Once he's gotten the plane up through the initial ascent, the pilot often powers down the acceleration and levels off before climbing again. This creates a sinking sensation in the stomach, and you may feel as though the plane actually is descending. But rest assured, it's just in a holding pattern—most likely to keep noise pollution down over urban areas or to make a sharp turn onto the correct flight path. If you're in a wing seat, you'll encounter some nervous-making sounds (clunking, whirring) prior to takeoff and during the flight. All the controls and devices on the airplane are either electrically or hydraulically activated. You may hear pumps that cycle on and off; they are designed to do that to maintain a certain pressure. (Adapted from www.anxieties.com.)

CURE-YOURSELF COURSES:

Most 12-step-style rehabilitations include education, breathing exercises, and suggestions for how to distract yourself. If none of that works, try prayer.

→**Fear-of-Flying-Help Course:** Designed by an airline pilot, this program includes photos, videos, sounds, and virtual-reality tours, if you can stomach them. *www.fearofflyinghelp.com.*

→**Fearless-Flight Kit:** Real-life stories, a personal checklist to bring onboard, and a Flight Harmonizer CD of comforting prose, music, and poetry. *www.fearless-flight.com*

→**Fear-of-Flying Clinic:** A behavioral counselor teaches you deep-breathing exercises so you don't die of asphyxiation on your way up. *www.fofc.com.*

For those of you who get more comfort out of knowing the statistics than from breathing exercises, you can browse "recent fatal events" involving airplanes at *www.AirSafe.com.*

BOZOPHOBIA AKA Coulrophobia: Fear of Clowns

Fun with Bo Bo and KoKo? Not on this playground. Whiteface, red noses, fright wigs—the reason people are scared of clowns is because they are really, really creepy. Think John Wayne Gacy—entertainer of hospitalized children in circus regalia by day, rapacious killer by night. The clowns of our day have turned from silly to sinister. They are two-faced and demented. Never trust a permasmile. And what other circus freak has inspired an entire genre of horror films? *Killer Klowns from Outer Space, Clownhouse, Killjoy, Stephen King's It, Fear of Clowns, S.I.C.K., Dead Clowns*—that's just the short list.

How can you, coulrophobe, cure your fear? A historical perspective helps. When you get scared, think of all the good clowns of yore. From Roman farce through the Harlequin of Commedia dell'Arte and Shakespeare's beguiling fools and court jesters, to the sad-sack Charlie Chaplin– and Buster Keaton–style tramps of early Hollywood, Bozo has had a long and venerable run. That poor old jester was every bit as scared as you are, his buffoonery a thin veil for prophetic wisdom that, upon its delivery, often raised the ire of his hot-tempered lord.

PHOBOPHOBIA: Fear of Fear

This is a hard-to-conquer problem. If one suffers from anxiety attacks (heart palpitations, cold sweats, chest pains), the very thought of an attack can cause one. It's a Möbius Strip of dread that can turn people into nonfunctioning blobs. Here are a few physical exercises that will help calm your nerves when the threat of becoming totally unhinged sets in:

→**Deep breaths:**
Inhale slowly for ten seconds, hold the air in your lungs for eight seconds, then exhale for another ten seconds. This will increase your oxygen intake and slow your heart rate.

→**Muscle curls:**
Flex your calf muscles for five seconds, then release them. Now flex your quads for five seconds and release them. Repeat the exercise throughout your body, isolating your musculature, squeezing, then releasing. This should have a relaxing effect. At the very least you're burning calories.

TECATE
GENUINE DRAFT

BEER-CAN ROOM DIVIDER

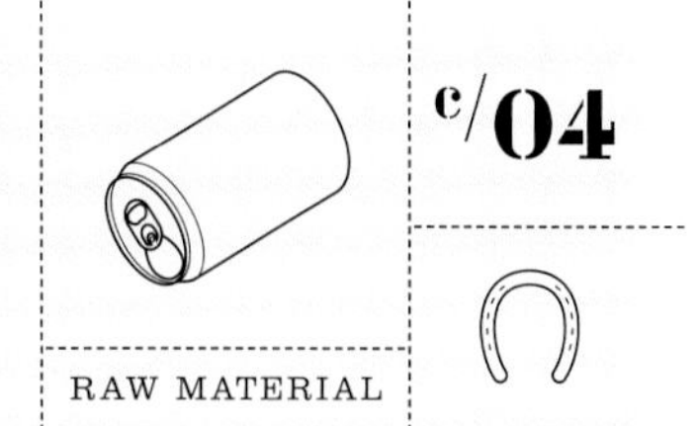

Forget 100 bottles of beer on the wall—we've got 216 cans that *are* the wall! A towering stack of empties can be more than a trophy or target-practice—use it to subdivide an open loft, screen unsightly exercise equipment, or hide the beer gut you built up in the process of emptying all those six-packs. And, utility aside, this aluminum and wooden-dowel shoji is a fun piece of pop art posing as furniture. So get to it—bottoms up! But do us the favor of getting good and sober before you start playing around with tin snips.

Year Peter Durance patented the use of metal containers for canning: 1810

Year Andy Warhol showed thirty-two Campbell's Soup cans in a Los Angeles gallery: 1962

BEER-CAN ROOM DIVIDER

$175

INGREDIENTS

+ 216 empty 12-ounce cans
+ 13 6' long foam pipe insulation 1/2" copper size
+ 12 6' long foam pipe insulation 1" copper size
+ 18 4 1/2' long 3/4" wood dowels
+ 32' of 16-gauge steel wire
+ 3' of 2" galvanized floor flange with 2" by 4" steel-pipe nipples (inside bottom can)
+ 18 "caps" for cans (2" flat cap hubs for plumbing)

TOOLS

+ Metal shears
+ Ruler
+ Wire cutters
+ Hacksaw
+ Pliers
+ Scissors
+ Drill
+ Hot glue gun

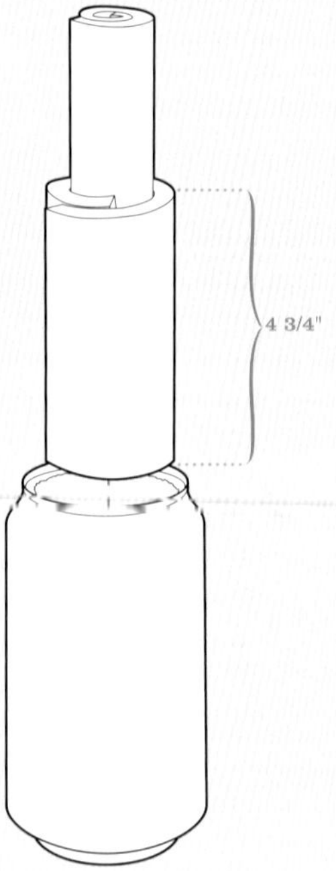

4 3/4"

(Figure A)

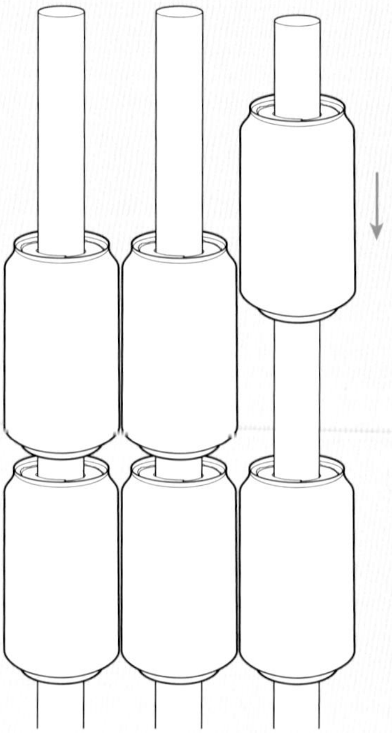

(Figure B)

1___ Use your metal shears to stab and cut away a 3/4" hole in the centers of the bottoms of the cans.

2___ Cut the entire bottom away from three of the cans (to hold pipe for feet).

3___ Use your metal shears to cut away the tops of all the cans.

4___ Stuff the smaller pipe insulation into the larger pipe insulation and trim into 4 3/4" pieces. You should net almost seventeen pieces for each 6'.

5___ Insert the insulation foam into the topless cans. You'll now have to separate the larger from the smaller, inserting the larger first, then the smaller. Do not attempt to remove the foam once it's inserted—those edges are sharp and jagged (Figure A).

6___ Plan and create your can design on paper, then thread twelve cans onto each dowel. Put the feet (the three cans from step 2) at the bottom of the first, middle, and last poles (Figure B).

7___ Drill holes through each dowel between (from bottom) the first and second cans, the fourth and fifth cans, the eighth and ninth cans, and the

Length of the longest steel bridge in the world (located in Japan): 4000m

She'd heard that magnets had healing powers, redirecting the way energy flows through the body. Feeling not herself, she strapped one to each wrist and each ankle before going to work. She wasn't sure she felt any different, but by the end of the day she'd collected ten paperclips, three pushpins, and a spoon.

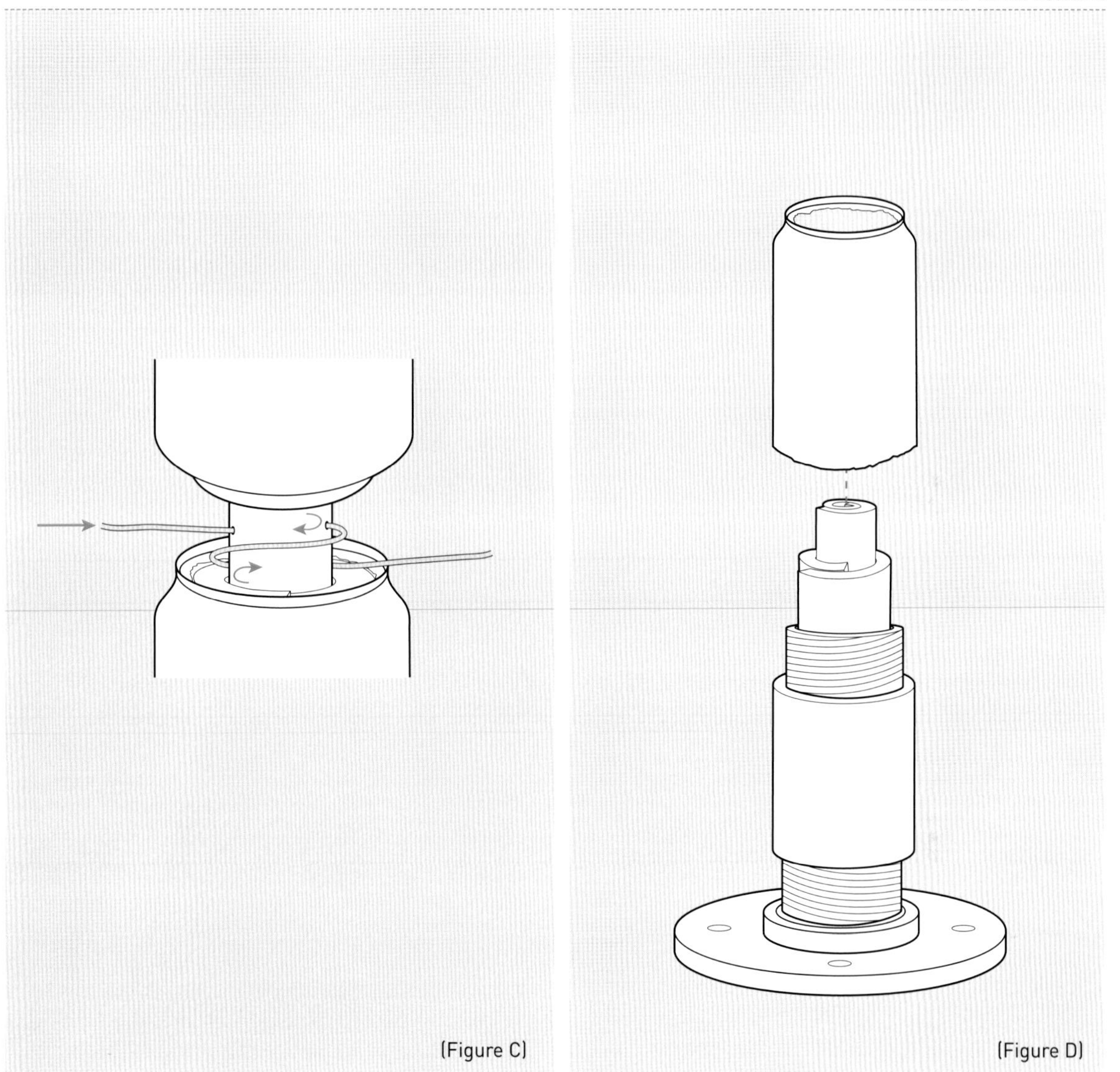
(Figure C)

(Figure D)

eleventh and twelfth cans. These holes are for the wire that will hold the can-poles together.

8 Lay the can-poles down on the floor and weave them together with wire. Thread the wire through the holes, wrap it around the pole again, twist, and repeat. Use pliers to pull the wire taut and keep it close to the poles so that the cans will camouflage it (Figure C).

9 To make the feet, screw 2" by 4" pipe nipples into 2" flange (Figure D).

10 Insert the pipe-insulating foam into the pipe nipples, as you did with the cans.

11 Take the cans you cut as feet and put them over the 4" pipe nipples.

12 Put the feet onto the first, middle, and last poles.

13 Raise your structure. Tighten the wires if necessary by twisting them.

14 Use a hot glue gun to affix the caps to the top row of cans.

15 Hide behind your wall of armor.

Tons of steel used in the towers, cables, and trusses of the bridge: 193,000

ART
cream
Artists & Prints
Viva Picasso
FRUITS
THE PHOTO BOOK
GREAT CAR
EARTH FROM ABOVE 365 DAYS
甘泉堂
DAYTONA
uncoated chips
LADDERS
LADDERS

LADDER SHELVING

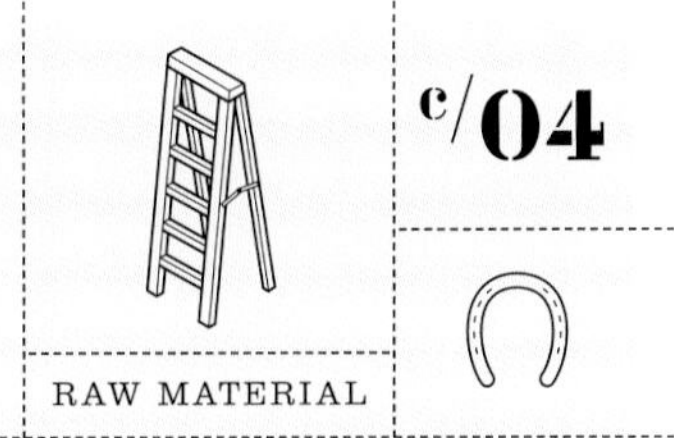

If you suffer from the lifelong problem of having too many books and too little wall space, this project is for you. Though we'd all like to own those built-in, wall-to-wall bookshelves we've seen in the fire-lit libraries of old movies—their rolling ladders giving access to the loftiest, most elevated tomes—we'll settle for this less-ambitious version. All that's required are two ladders (if you can find the old wooden variety, all the better), and a few sturdy sheets of plywood. Almost no assembly required.

The brothers stood at the highest point of the land their father had left them, looking down over its sun-bathed valleys, searching for a name. They saw for the first time how the river snaked in, then turned back on itself, as though it had changed its mind. From then on, they called it horseshoe ranch, and had great luck thereafter.

LADDER SHELVING

$120

INGREDIENTS

+ 2 6' folding aluminum ladders
+ 5 shelves in varying widths and a standard 83" in length
+ 85' of iron-on wood veneer tape
+ Semi-gloss polyurethane

(Note: The total wood needed for this project is 45 1/2 square feet.)

TOOLS

+ Ruler
+ Hand circular saw
+ Iron
+ Scissors
+ Sandpaper
+ Paintbrush

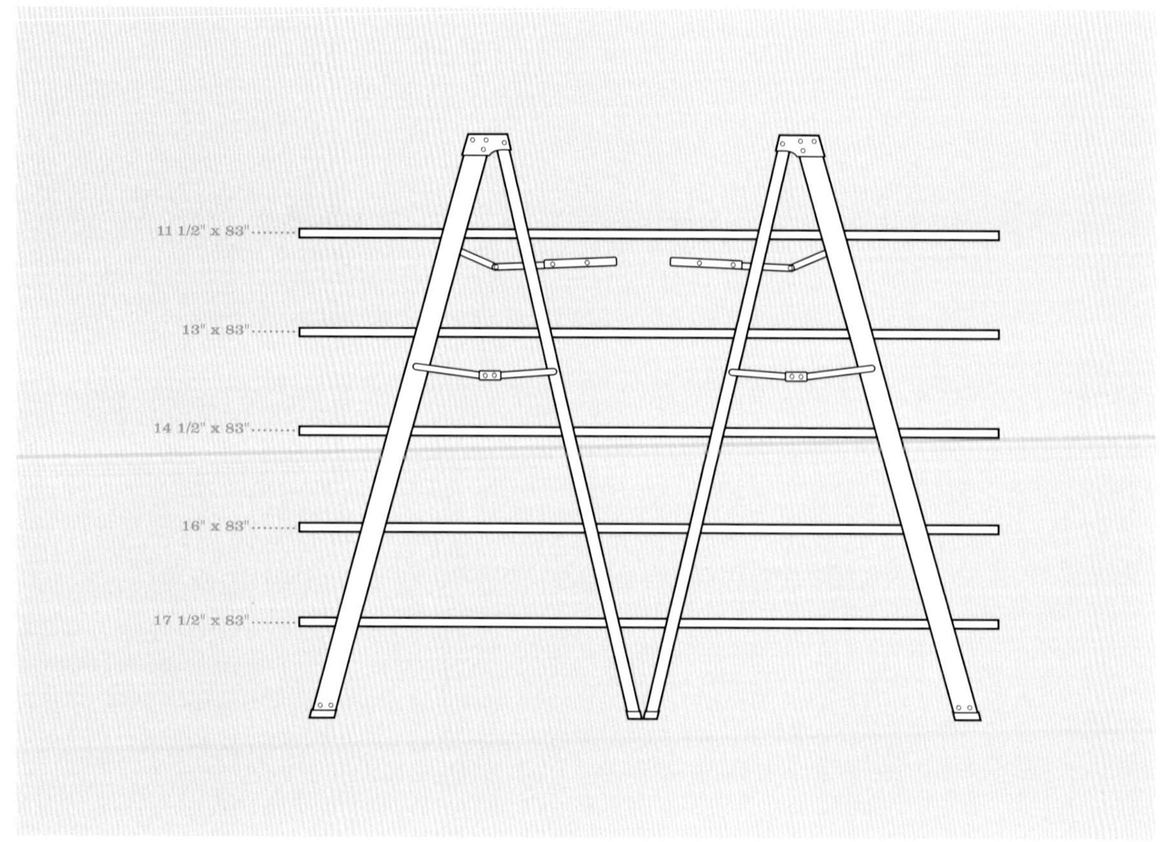

1 Stand your ladders next to each other, with the steps facing out.

2 Measure, mark, and cut five shelves with your circular saw in the following dimensions—yes, the steps of a ladder narrow as you ascend.
11 1/2" by 83"
13" by 83"
14 1/2" by 83"
16" by 83"
17 1/2" by 83"

3 Iron the veneer tape onto the edges of your wood shelves.

4 Trim and sand any excess tape.

5 Brush two coats of polyurethane over the tops and sides of your shelves.

6 Let dry.

7 Rack up your shelves on the ladders. (You may have to get a metal file and file a little bit of the underside of the ladder shelf if the wood planks are too tight of a fit.)

8 Move the assembly into place, and stack it up with books and other keepsakes.

Length of the "steel dragon," the worlds longest roller coaster: 2.5 kilometers

Temperature at which steel pipes remain impermeable: 580° C

IRON MAN

A LOOK BACK AT THE ORIGINS OF HEAVY METAL

YOU MAY *THINK* you don't need to hear this, but if you have even a passing interest in rock 'n' roll, then listen up. It all started in the late 1960s in Birmingham, England, where long-haired British lads became obsessed with American blues music and the psychedelic sounds associated with the Fillmore West in San Francisco (Jimi Hendrix, the Doors, Janis Joplin, Jefferson Airplane). The tribute is obvious in Mick Jagger's hip-swiveling, slack-jawed slur, but it was the Yardbirds (Jeff Beck, Eric Clapton, Jimmy Page), Cream (Eric Clapton really got around), Led Zeppelin (which Jimmy Page founded after leaving the Yardbirds), and, most importantly (music swells), Black Sabbath, fronted by tongue-waggling sitcom star Ozzy Osbourne, who took the volume up to 11. They started out covering blues standards but replaced the acoustic guitar with an electric ax. They also introduced some dissonance and made liberal use of the booming power chords that the Kinks and the Who were already playing. Once the guitars got loud, the singers and drummers had to ratchet up their parts, too. What resulted was an amplified, dirtier, more urgent sound—the blues on amphetamines.

By the mid '70s, a Long Island quartet called Kiss, taking cues from nearby glam bands like the New York Dolls, were performing pyrotechnic, eardrum-popping shows in full makeup and regalia. Kiss would come to define American heavy rock for much of the decade, until it began to cross-breed with early punk (cf. the Sex Pistols, whose manager Malcolm McLaren was also influenced by the Dolls) and an era of harder, faster music began. Back in the UK, a new wave of British metal was born, largely in response to the arena-rock commercialization of first-gen groups like Led Zeppelin and Deep Purple, with acts like Judas Priest and Iron Maiden rising to prominence, eventually supplanting their bloated forebears as the headliners of large-scale, lighters-aloft world tours. The association with African-American blues was all but forgotten in the American hair-metal scene, which took off in LA in the early '80s. Mötley Crüe, Guns N' Roses, Poison, and Ratt found all the inspiration they needed in tight spandex and painted faces, trading licks on MTV as America's suburban white kids sat agog, growing out their mullets.

HAIR METAL WAS THE LOGICAL CONCLUSION OF BOWIE'S SEXUALLY AMBIGUOUS GLITTER ROCK BUT WITH MORE MACHISMO.

Hair metal was the logical conclusion of Bowie's sexually ambiguous glitter rock, but with more machismo. Into the mid-'80s and '90s, louder, more extreme offshoots of heavy metal like thrash, death metal, black metal, and grindcore began attracting alienated youths in London, New York, San Francisco, and Tampa, Florida, where groups with evocative names like Slayer, Anthrax, Napalm Death, and Morbid Angel brought the music to new levels of parental offensiveness. The Dungeons & Dragons themes of darkness, evil, sexual domination, and apocalypse held irresistible appeal for kids who needed a higher daily intake of anger than bubblegum pop or new wave could provide. It was a natural, if geometric, evolution from the "Helter Skelter" era of the Beatles, who had dropped the sunny, all-you-need-is-love vibe that suited them so well in the '60s for an edgier sound, one better suited to the era of Vietnam and Watergate.

So how did a term applied to the elements at the center of the periodic table end up describing headbanger music? Our personal opinion is that it came from Jimmy Page, who changed the spelling of his group's name from Lead Zeppelin. As the story goes, Jimmy Page had put together a band called the New Yardbirds to make good on some tour dates in Europe that the by-then mostly dispersed group (i.e., Eric Clapton and Jeff Beck) had booked. Keith Moon, drummer for the Who, is reported to have said, "With that lineup you'll go down like a lead balloon," referring to the original lead-constructed zeppelin and its calamitous crash. Ironic gestures ruled the day, and Page promptly renamed the band, changing the spelling from "lead" to Led so we Americans wouldn't pronounce it with a hard E, as in "lead singer."

But there are as many theories as to the source of the phrase as there are rock critics to float them. Foundational myths include William Burroughs's 1962 novel *The Soft Machine*, where he introduced Uranian Willy, the Heavy Metal Kid, and the subsequent *Nova Express* (1964), in which the term was used as a metaphor for drugs: "With their diseases and orgasm drugs and their sexless parasite life forms—Heavy Metal People of Uranus wrapped in cool blue mist of vaporized bank notes—and the Insect People of Minraud with metal music . . ."

Also, Birmingham, the city that many of the forefathers of metal called home (Zeppelin, Sabbath, the Move), was a center of industry, and the rockers were no doubt breathing in steel- and iron-ore particulate matter throughout their childhoods. Biographers of the Move write that the band's "heavy" guitar riffs were popular among those in the "metal midlands."

The first appearance of "heavy metal" in a song lyric is in the 1968 Steppenwolf song "Born to Be Wild":

I like smoke and lightning
Heavy metal thunder
Racin' with the wind
And the feelin' that I'm under

The first well-documented use of the phrase to refer to a style of music appears to be the May 1971 issue of *Creem*, in a review of Sir Lord Baltimore's *Kingdom Come*. The key phrase: "Sir Lord Baltimore seems to have down pat most all the best heavy metal tricks in the book."

HOW TO GET HEAVY

1. Add a diacritical mark to your name. An O or A should take an umlaut. If you have no O's or A's in your name, consider changing it to something that sounds German or Swedish.
2. Grow your hair long in the back, but leave it shaggy and tousled in the front. If that doesn't work for you, get a perm.
3. Drive a Trans Am or Firebird and airbrush flames or lightning bolts onto the hood.
4. Watch *Spinal Tap* repeatedly and try to pick up on the accent and apparel.
5. Practice flipping your hair back and forth to the beat. Repeat.

ZEPPELIN ROCKS

HEAVY METAL BUYER'S GUIDE

ESSENTIAL LIBRARY

1. Jimi Hendrix Experience: *Are You Experienced?*
2. Led Zeppelin: *Physical Graffiti*
3. The Beatles: *The White Album*
4. New York Dolls: *New York Dolls*
5. Black Sabbath: *Paranoid*
6. Kiss: *Destroyer*
7. Iron Maiden: *Number of the Beast*
8. AC/DC: *Back in Black*
9. Mötorhead: *Overkill*
10. Scorpions: *Blackout*
11. Slayer: *Reign in Blood*
12. Judas Priest: *Screaming for Vengeance*
13. Van Halen: *Van Halen*
14. Metallica: *Kill 'Em All*

HAIR METAL STARTER KIT

1. Poison: *Open Up & Say . . . Ahh!*
2. Whitesnake: *Whitesnake*
3. Ratt: *Out of the Cellar*
4. Def Leppard: *Pyromania*
5. Mötley Crüe: *Shout at the Devil*

DUCHAMP'S CORNER

16 Alternate Uses for Paper Clips

X-mas tree ornament hooks
Zipper pull
Key ring
Engagement ring
Lock breaker
Glue unplugger
Necklace
Poker currency
Tie clip
Money clip
Hole poker
Concrete etcher
Easter-egg dipper
Hinge saver
Fishing hook
Cord manager

SLAG HEAP

Where to find scrap metal

Junk Yard Dog *(www.junkyarddog.com)* Auto-part-salvage haven for all of your hubcap fountain and windshield fire-screen needs.

Knight Replicas *(www.knightreplicas.com/how_to_find_parts_at_salvage_y.html)* Good advice on where to find your hubcaps and siding. Tips on finding salvage near you using Yahoo, what to wear on your Dumpster dive, and "the sacred rules" of hunting for parts at scrap yards.

c/04 ☐ Su ☐ Mo ☐ Tu ☐ We ☐ Th ☐ Fr ☐ Sa

Why Don't I Do This Every Day?

TESTIMONIAL # 0037

NAME: **Sean Cooper**

FROM: **Brooklyn, New York**

OCCUPATION: **Unemployed**

Sean, why don't you do these things every day?

The world is so full of things, the fewer I have of them the better. Except, no, that's a lie I suppose. I just bought a new couch. Why didn't I make it out of toothpicks and junk mail and bits of ficus clippings? I guess because then I'd have a couch made out of toothpicks and junk mail and bits of ficus clippings. Which, I mean, is okay I guess, if you're, you know, into that sort of thing, toothpicks and junk mail . . . and bits of ficus clippings.

1

TOOL FURNITURE

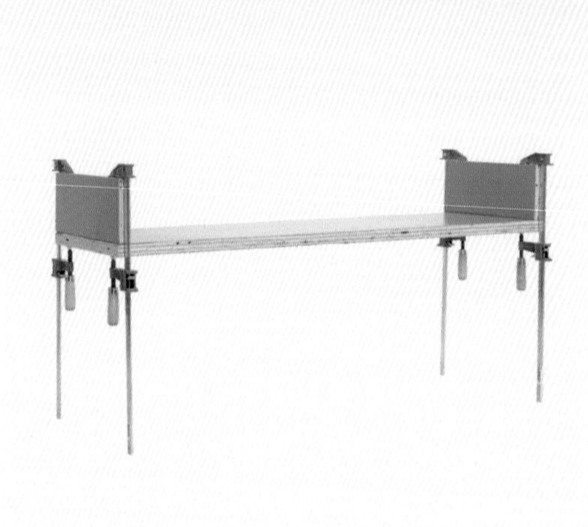

INGREDIENTS

+ 2 pieces of 1" by 5" by 15" plywood
+ 1 piece of 1" by 18" by 48" plywood
+ Wood stain or paint
+ 4 22" bar clamps

TOOLS

+ Sandpaper
+ Paint brush

01 Lightly sand the wood edges and the faces of your boards for sitable smoothness.

02 Varnish the wood pieces and allow them to dry.

03 Clamp a 5" by 18" plywood piece at a 90 degree angle to the edge of the 18" by 48" piece of wood.

04 Repeat step 3 for the other side.

05 Stand it up on all four "legs."

06 Sit down and give your feet a rest.

Why don't we?

2

BBQ POT RACK

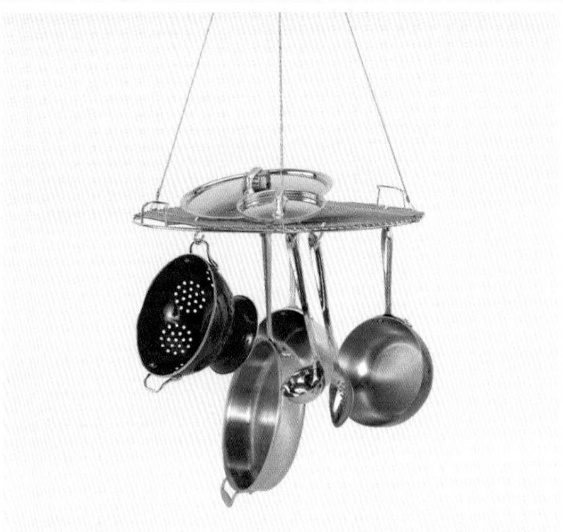

INGREDIENTS	TOOLS
+ Barbecue-grill rack	+ Hand drill
+ Screw eye	+ Small drill bit
+ 3 to 4 yards of 1/16" cable	+ Hammer
+ 8 cable ferrules	
+ Lap link	
+ 12 S-hooks	

01 Decide where you want to hang your pot rack. Pick a place that will provide you convenience of use but won't create frequent opportunities for blows to the head.

02 Insert your screw eye into the ceiling. Make sure you mount it securely—falling pots are dangerous.

03 Connect a piece of cable to the grill by hammering a ferrule to secure the cable onto four corners of the rack.

04 Cut your cable into four equal pieces. Create a closed loop by hammering a ferrule at the ends of each of the four cable lengths.

05 Hook the four closed looped ends onto the lap link.

06 Hook your lap link to the ceiling with the screw eye.

07 Place the S-hooks on the grill to hang pots and pans and large cooking utensils.

08 Hang your cookware.

We really should, you know.

3

COLANDER LIGHT SCONCE

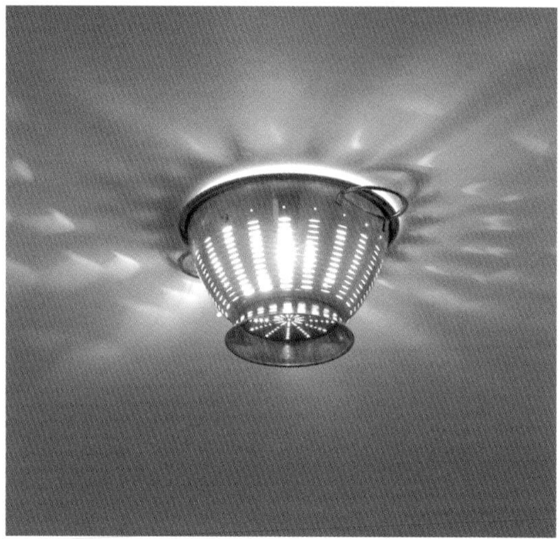

INGREDIENTS	TOOLS
+ One large stainless-steel colander (the 3-quart size is good for a single bulb)	+ Power drill
+ Drywall screws	
+ 2 one-hole conduit straps	

01 Locate a live, exposed bulb in your ceiling or wall.

02 Use drywall screws to secure the colander to the wall through the conduit straps on the handles. (You can also use picture-hanging hardware to hang the collander on the wall.)

03 Turn it on and check out the shadows.

Tarry no longer, chop chop!

Chapter: 05

GLASS

HALF FULL

There are many inventions that still puzzle us, like how setting a needle in a groove in vinyl produces music, how lenses and mirrors affix images to film, and how stuffing a duck inside a turkey makes for a popular Thanksgiving meal. Another fact that never fails to amaze: Glass is manufactured from common sand. By fortuitous fluke, Phoenician traders enjoying a midroute meal set up their fires on riverbanks and used blocks of nitrate to support their cooking pots. When the nitrate melted and mixed with the sand, it produced a clear liquid that dried hard and semi-transparent.

Jump forward seven thousand years or so and look how dependent upon glass we've become. Without it we'd still be drinking our wine from silver goblets, walled inside our homes with no view of what lay beyond, blind to the far reaches of the heavens and the microscopic depths of matter. Glass permeates our language, too. Careers are kept in check by glass ceilings. A person's outlook is gauged by whether he sees the glass as half full or half empty. The epigrammatic warning to hypocrites is this: Those who live in glass houses should not throw stones. When a material achieves the status of metaphor (steely demeanor, concrete jungle), it has officially arrived.

Anything with glass's ubiquitous appeal is bound to spin off an entire product line. We now have tempered safety glass, used in windshields and shower doors (the car and bathroom being two places where slippery-when-wet accidents are likely to happen); wired glass, embedded with a steel mesh to keep the pieces in place should breakage occur; and super-friable stunt glass, used to convey dramatic realism on stage (for instance, breaking a bottle over the bad guy's head). There are also scores of decorative styles that have developed over the years, many of them by Louis Comfort Tiffany, founder of the Fifth Avenue destination Tiffany & Co. Among them: crystal cuts that give glass the appearance of diamonds; beveled edges; hand-blown or -rolled pieces that crackle or show interior bubbles; frosted or opalescent "drapery" glass that has a milky cast; the stained variety seen in chapels worldwide; and water glass that has a wavy, rippled effect. Our friend Nikolas Weinstein made a hand-blown glass chandelier that's shaped to look like the rivulets in bacon. His PP3 Chandelier is located in the central atrium of the Frank Gehry–designed DZ Bank, on historic Pariser Platz in Berlin. An airborne constellation of thirty-four glass panels made of fused-together tubes that are then slumped over a mold, the installation covers more than two thousand square feet and weighs 2.5 tons.

Without it we'd still be drinking our wine from animal skeins or silver goblets.

Though we have no glass-blowing furnace to make such a thing, castoffs are easy to come by, and produced without any toxic by-products that might make you sick in the process of reusing them. Glass may be the most innocuous building material after mud. Here are a few half-full ideas for what to do with yours.

Construction:
Leger Wanaselja Architecture

A CLOSER LOOK AT THE HISTORY OF GLASS

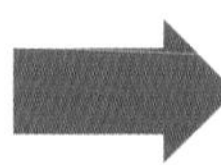

2500
Egyptian glass beads date back roughly to this time.

650
Foreseeing how important glass could become, Assyrians inscribe glass making instructions in stone tablets for future generations.

1200–1300
Venice becomes the center of glassmaking in Europe. Murano, Italy, islanders develop soda lime, and Venetians term this clear and thin glass *cristallo*, which is still in use today.

1500s
Glass is silvered to make it more reflective. Such mirrors—beyond their importance to the aristocracy, who could now gaze at themselves adoringly—greatly aided astronomers in looking more closely at the heavens. With his self-crafted telescopes, Galileo claimed to have seen mountains on the moon, to discern that the Milky Way was made up of tiny stars, and to have seen four small moons orbiting Jupiter.

Early 1800s
The demand for window glass rises with its mass-production during the industrial revolution.

1904
A glut of glass bottles and jars flood the market as Michael Owen patents his glass-shaping machine.

1930
Sunglasses become popular just in time for the uptown hipsters of the Harlem Renaissance.

1950s
Sir Alastair Pilkington introduces a manufacturing method called float glass production, by which 90 percent of flat glass is still manufactured today.

The hard shiny stuff starts with

petrified lightning, and ends with

curbside recycling.

Beginning of time
There are two forms of natural glass that have been used since humankind knew what to do with itself. When lightning strikes sand, the heat sometimes fuses the grains into long, slender tubes called fulgurites, or petrified lightning. Also, a volcanic eruption can fuse rocks and sand into obsidian, which early civilizations shaped into knives, arrowheads, and jewelry.

5000 BC
The ingredients that make up glass are discovered accidentally when merchants place their cooking pots on blocks of nitrate by the fire. The intense heat melts the blocks and mixes with the sand of the beach, forming an opaque liquid, which then cools into glass.

AD 100
Glassblowing begins. The Syrians create the first vision aid, a glass sphere, and call it a "reading stone," i.e., a magnifying glass. They pass the knowledge along to the Romans.

1100s and 1200s
Stained-glass windows reach their apex of popularity in Europe, making churches more colorful.

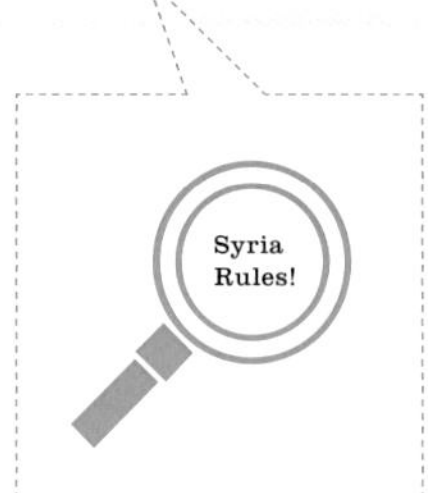

1240
Italians invent first wearable eyeglasses, to be used by literati like Dante Alighieri.

1858
The screw-top mason jar for home canning appears.

1880
Commercial food packers begin to use glass containers to preserve food. Glass tableware becomes popular.

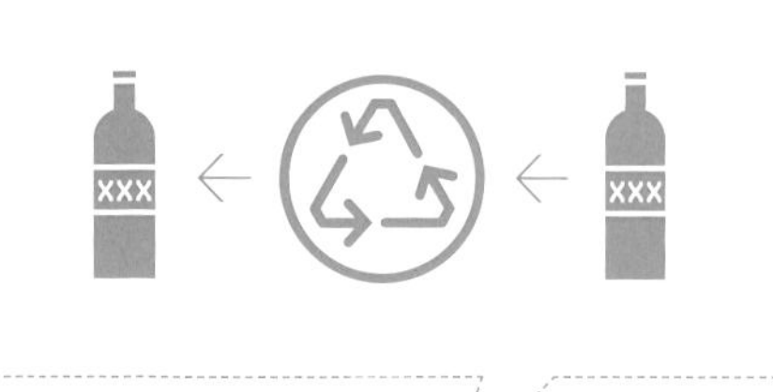

1902
Irving W. Colburn patents his sheet-glass drawing machine for the mass-production of windows.

1960s
Recycling gains popularity, requiring the establishment of collection centers where people can return their empty whisky and wine bottles. A decade of guilt-free drinking ensues.

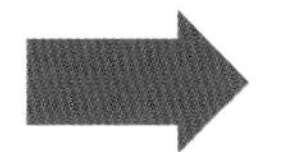

CHEMICAL BREAKDOWN

GLASS

SAND

SODIUM CARBONATE

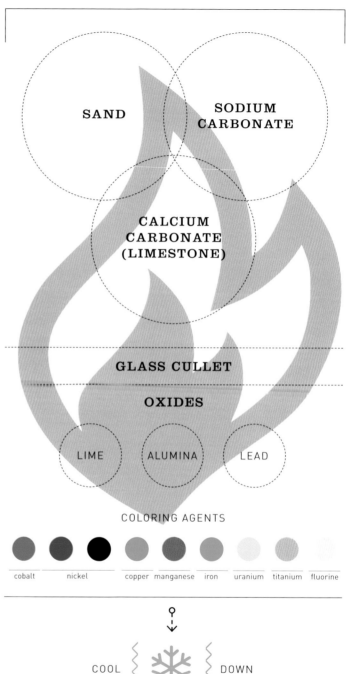

CALCIUM CARBONATE (LIMESTONE)

GLASS CULLET

OXIDES

LIME

ALUMINA

LEAD

COLORING AGENTS

cobalt | nickel | copper | manganese | iron | uranium | titanium | fluorine

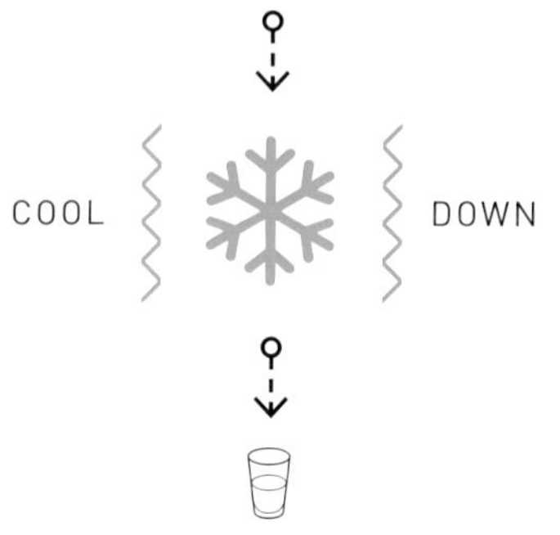

01

Glass is made from a mixture of sand, sodium carbonate, and calcium carbonate (limestone) that is heated to high temperatures and then cooled. In addition to these raw materials, glass *cullet* (a gorgeous word for the broken waste glass returned from a recycling facility) should be added to the mix to accelerate melting and increase the strength of the glass. The ideal grain size to make glass is between 0.1 and 0.4 mm, and it should be as colorless as possible. Though almost half of the Earth's land surface is sand, most of it contains impurities and colorings that can't be used for glassmaking. The purest sands, unsurprisingly, are found in Australia, Indonesia, Malaysia, and Vietnam.

02

Soda ash (or sodium carbonate) or glauber's salt (sodium sulfate) and potash (potassium carbonate) are added as a white powder. When the batch is melted down, sodium oxide becomes part of the glass, and carbon dioxide is released as a by-product. If no coloring materials are added, this produces colorless glass.

03

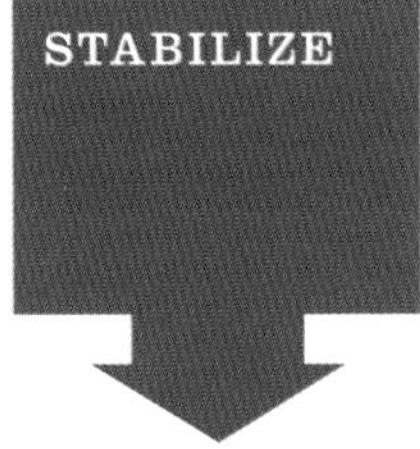

Oxides can be added to the batch to give the glass certain physical and chemical properties. The addition of lime (found in nature as limestone, marble, or chalk) improves the output's hardness and chemical resistance; alumina (aluminum oxide) also increases chemical resistance. Lead oxide will decrease hardness and make the finished product look more reflective and glassy.

04

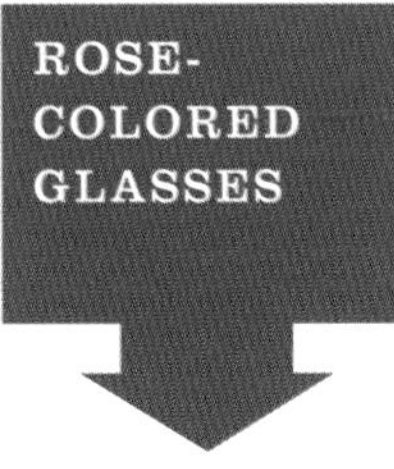

Only pure chemicals are used as glass-coloring agents. Adding metal oxides like nickel (blue, violet, or black) copper (turquoise), manganese (decolorizes and, at high concentrations, adds an amethyst tint), iron (green), cobalt (blue), uranium (fluorescent yellow or green, plus it's radioactive!), or titanium (ocher) will change the color of glass. When fluorine-containing materials are added, small crystalline particles form, giving the glass an opaque, milky look that's used by the lighting industry to head off bulb glare.

05

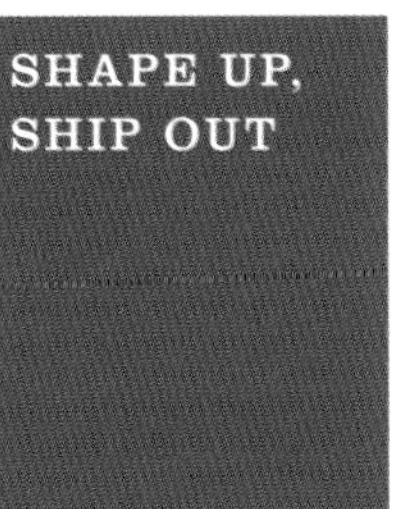

The hot molten glass is hand-blown or poured into molds, set to dry, then wrapped well for the bumpy ride to its next stop.

SECESSION Wien
SECESSION Wien

WATER-BOTTLE CHANDELIER

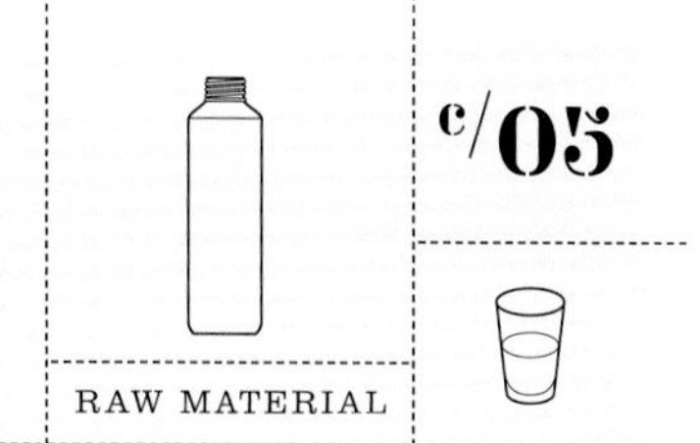

No Bohemian crystal gewgaws for us. These VOS carafes, emptied of their boutique Norwegian spring water by those who'd shudder to think of drinking tap, have more snob appeal. We used a bottle cutter to lop off the bottoms (it's our new favorite tool; get one at your local hobby shop or art store), then strung them with Ikea pendant-lamp sockets. Unlike their four-figure cousins, the poor man's crystal is everywhere, and free. (We asked an upscale eatery for some of their recycling and they gladly obliged.) Vary the height of your VOS like hanging icicles, or line them up military style.

Millions of glass bottles and jars thrown away by Americans every year: 27

How many years later glass relics are found in perfect condition: 4,000

WATER-BOTTLE CHANDELIER

$95

INGREDIENTS

+ 9 glass bottles (We found our cylindrical ones at a local restaurant. Any glass bottle that accommodates the diameter of your bulb and fixture will work.)
+ 9 pre-wired light receptacles (ours were Hemma, from Ikea)
+ 9 lightbulbs
+ Electrical tape
+ 2 power strips
+ Large ceiling eye hook (to mount overhead)
+ Plastic tie

TOOLS

+ Bottle-cutter kit (runs about $50 at a craft or hobby shop)
+ Wire cutter
+ Wire stripper
+ Glass plate (you will use this to sand down your cut glass)
+ Candle
+ Matches
+ Ice cubes
+ Kitchen towel

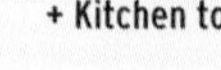

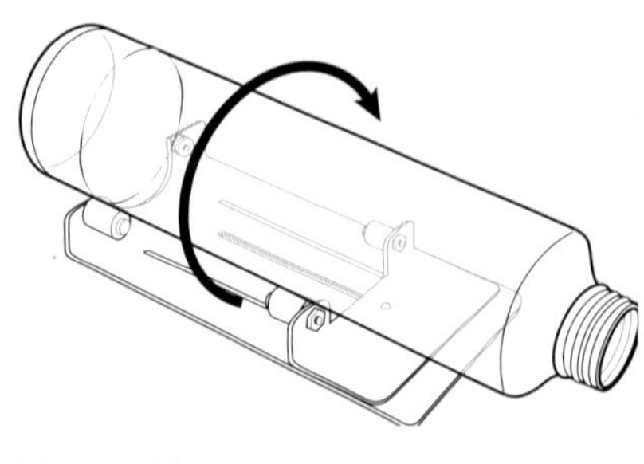

FIGURE A

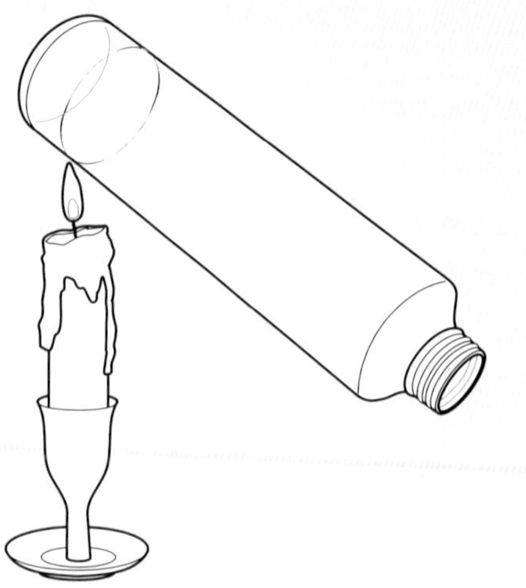

FIGURE B

1 Remove caps, rinse out the bottles, and let them air dry.

2 Set up the bottle cutter and determine how long you want your pendant to be. Make sure you have a solid end guide for the bottle when you rotate it on the cutter so your etch line will connect. We used a small, heavy box.

3 Using both hands, rotate a bottle in one continuous motion, applying even pressure throughout the rotation for best result (Figure A).

4 Light the candle. Remove the bottle from the cutter and hold the etch line above the lit candle to heat the cut.

5 Slowly rotate the etch line above the flame in a continuous motion to apply even heat. If it gets too hot, you will hear the bottle crack. If you don't heat it up enough, the glass won't break off as described in the next two steps. This is a trial-and-error part of the process that depends on the thickness of your glass (Figure B).

6 Remove the bottle from the candle heat and apply an ice cube directly along the circumference of the etch line, letting it drip over the towel (Figure C).

7 The end of your bottle should break off by itself from the change in temperature. If not, using the towel, hold the end that needs to break off and gently twist. It should come off easily. (If it doesn't, dry off the bottle and repeat steps 5 through 7.)

8 Wipe the end of the cut bottle and set aside.

9 Repeat steps 3 through 8 for each of your bottles.

10 Set your small glass plate

Number of hours a lightbulb will remain lit from the energy saved by recycling one glass bottle or jar: 4

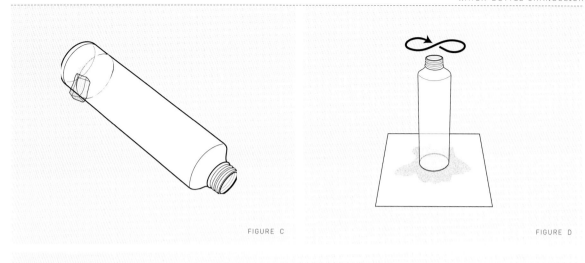

FIGURE C

FIGURE D

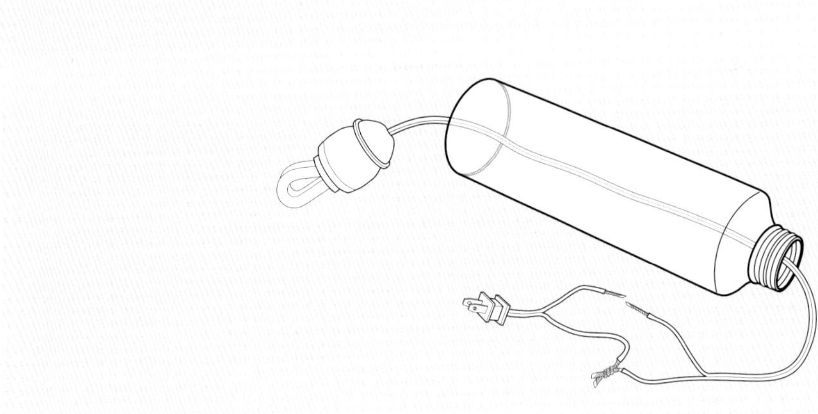

FIGURE E

on a level surface and sprinkle about a teaspoon of the carbide polishing powder that came with your kit and $^{1}/_{4}$ teaspoon of water onto the glass plate.

11 Set the cut side of one of the bottles down on the glass plate.

12 Rotate the cut side to the flat of the glass in a figure-8 motion. The polishing powder and the water, combined with the rotating grind, sands the cut glass into a smoother edge. This process is very loud and will take about three to five minutes per bottle (Figure D).

13 Wipe the bottle edge clean and set aside.

14 Repeat steps 11 through 13 for all the bottles.

15 Because the plug end of the light socket will not fit through the mouth of the bottle, you will have to cut the wire (preferrably closer to the plug end).

16 Feed the cut end attached to the socket through the bottom of the glass pendant and up through the bottle's mouth (Figure E).

17 Next, strip both cut ends of the wire, reattach, and wrap with electrical tape for a secure, safe connection.

18 Plug it in to see if it works.

19 Repeat steps 15 through 18 for the rest of the bottles.

20 Once you have all the sockets threaded through the bottles, arrange, and use a twist tie to hold the electrical cords together.

21 Secure to your ceiling hook, plug in the chandelier, and enjoy a glass of high-class water.

Percentage less energy required to make glass from recycled cullet rather than from all new material: 30

PLEXI POST-IT BOARD

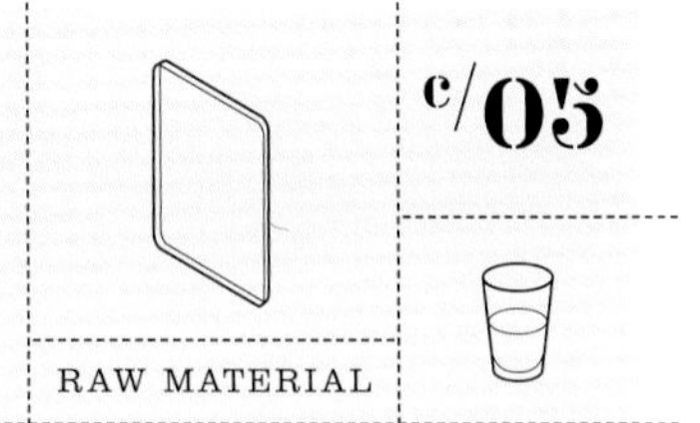

Do you have enough real estate devoted to to-do lists, postcards, and snapshots? Ours just seem to breed uncontrollably, ordering us to add more refrigerator doors, more cork-board, more wall space! Searching for alternatives, we made this Post-it panel with a plate of salvaged Plexiglas. (Sure we know it's plastic, not glass—it just fit better here.) Check your local plastics supplier for Plexi remnants and an art store for the large rubber bands used for sketch boards. Pulled taut, the elastics hold as many items as you can fit under them, and do away once and for all with pesky tack holes.

Percent recycling glass instead of making it from silica sand reduces mining waste: 70

PLEXI POST-IT BOARD

$39

INGREDIENTS

+ Piece of Plexiglas not to exceed 30" by 30" (so the rubber bands won't overstretch themselves)
+ 20 large rubber bands (used for portable sketch clipboards)
+ 2 D-rings and screws

TOOLS

+ Fine sandpaper
+ Clean rag
+ Window cleaner
+ Power drill with 1/16" drill bit
+ Phillips-head screwdriver

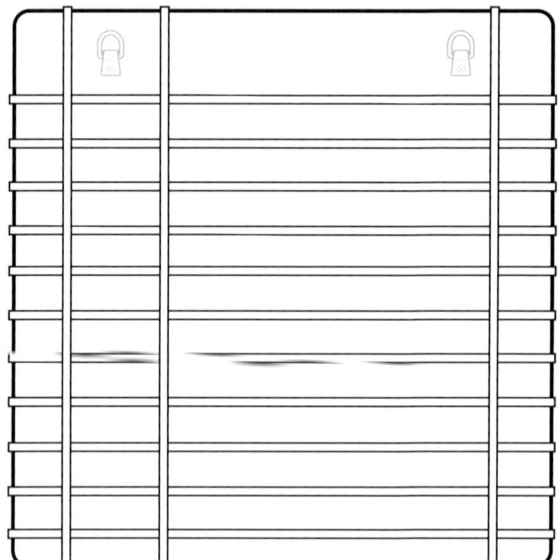

1 Lightly sand the edges of the Plexiglas if they are not already smooth. This will give your rubber bands a longer life.

2 Use a clean rag and window cleaner to remove any smudges from the Plexiglas.

3 Decide how you want your rubber bands arranged. Make sure to vary the distance between the bands to give your board versatility in accommodating a variety of board matter.

4 Stretch the rubber bands around the board.

5 Using the drill and 1/16" drill bit, pre-drill two small holes (not all the way through) into the Plexi where you will put the D-rings (one on each side). This prevents the board from cracking when you screw in the D-rings.

6 Screw in the D-rings on the upper right and left corners of the backside of your bulletin board.

7 Mount on the wall and fill with all the itty bits of paper on your desk.

Number of people employed by U.S. glass industry: **50,000**

They called them "fight plates"—a stack of cheap glass tableware that they kept on the sideboard at all times. Whenever a quarrel began spiraling out of control, one of them would take a plate and smash it, at which point they'd both start laughing, and the fight would be over.

WANT TO MAKE A DARK, BROODING, INDIE FILM? FOLLOW THIS HANDY FORMULA DEVELOPED BY INGMAR BERGMAN.

THROUGH A GLASS DARKLY

HOW TO MAKE A FILM IN THE ART-HOUSE STYLE

I came out of that movie house reeling like a drunkard, drugged speechless, with the film rushing through my bloodstream, *pumping and thudding.*
—Swedish actress Gunnel Lindblom, describing her first Bergman film experience in 1949

ART-HOUSE FILMS have their own formula. When you want to make cinema verité, there are tricks to capturing the gritty intimacies and true pacing of life on film without sending audiences walking. We take as our case study Ingmar Bergman's cheerless 1961 psychodrama *Through a Glass Darkly,* about a young woman's schizophrenia and eventual descent into madness. Harriet Andersson plays Karin, who has returned home to a characteristically bone-chilling island in Sweden after a stint in a psychiatric hospital. Though she recalls her carefree days there as a child, it doesn't take long for her to start coming unhinged in all the ways you'd expect from a beautiful young art-house starlet: She believes God to be a spider, is possessed by voices, and coaxes her brother into an incestuous relationship. All of this is observed and analyzed by Karin's emotionally remote father (played by Gunnar Björnstrand), who takes copious notes in his journal, documenting her behavior. Karin realizes she no longer loves her husband (the great, bony-faced Max von Sydow), who, like everyone else, is too freaked out by her behavior to help lift her out of her free fall.

The unswerving gaze of the art film upon the human psyche acts as a kind of chemical peel, sloughing off our game face to reveal the raw dermis of existence underneath. Groan away, but we'll watch an emotionally gnarly film over Hollywood fluff any day. Sure, a happy ending has its place, but when you want to feel better about your life, with its catalog of problems, turn to darker cinema—everyone's in worse shape than you, and they'll go on and on about it for two hours or more. If you've never seen a European film, we suggest that you introduce yourself slowly. Watching Bergman's moral fable *The Seventh Seal* will give newbies the aquavit nip of existential gloom within the first few minutes; tarry further and fatal depression could result. But if, like us, you become smitten with the genre—if, indeed, you cannot get enough of it, and quickly assemble a lengthy Netflix queue of cult films, only to decide that your life cannot be complete until you contribute your own sharply observed classic to that black pool of cinematic disconsolation—here are a few pointers on how to transform oppressive winters, family dysfunction, and drug-addled breakdowns into powerful, enduring art.

THE UNSWERVING GAZE OF THE ART FILM UPON THE HUMAN PSYCHE ACTS AS A KIND OF CHEMICAL PEEL, SLOUGHING OFF OUR GAME FACE TO REVEAL THE RAW DERMIS OF EXISTENCE UNDERNEATH.

HOW TO MAKE A FILM LIKE INGMAR BERGMAN: A CRIB SHEET

1 → Choose an ill-fated subject. Your script should have all the star-crossed doom of a Greek tragedy: two people who fall in love, only to discover that they're blood relatives; a famous actress stricken with a sudden case of stage fright that turns into a full-blown ontological crisis; a bad case of food poisoning that leads the hero to investigate industrial farming practices, eventually meeting his fate at the hands of corporate goons. With no shortage of real-life drama to draw upon, it's easy to take a tale of moderate hardship and whip it into a complete disaster. Dial up childhood memories, the evening news, or what you imagine about people at the office and you're off to the races!

2 → Shoot your film in black-and-white—nothing exaggerates the harsh dualities of life quite like it. Though the format can be soft and luminescent, as it is in Hollywood's romantic capers of the '30s and '40s, it can also be the cinematic equivalent of a migraine—severe, bled of color, at times blinding. Add a German score to the scene and you've got your audience staring into the void!

3 → Make sure there are plenty of long (sixty-second minimum) close-ups. The viewer should be virtually face-to-face with the character. This creates an almost visceral transfer of feeling through the screen. Look at those heavy-lidded eyes, the quivering lips—do you not feel her pain? These shots can involve mirrors, as in Bergman's famous transference scene in *Persona*, where Bibi Andersson's countenance morphs into Liv Ullman's. The mirror is a common cinematic prop, and it helps make staring blankly into the camera with nothing to say seem a little less autistic.

4 → The landscape should be wintry, with drawn-out pans of windswept fields and brute escarpments. Remember: A classic art-house film requires the Weather Channel drama of northern climes. You are creating a visual metaphor for sudden psychic plunges and ventures into the deepest recesses of the mind. No winter nearby? Scout a location in Nova Scotia.

5 → All of the characters should have great bone structure. Max von Sydow, the leading man of the Bergman films, has the kind of jowls that seem as though they're being pulled into the depths of hell by an invisible force. And Liv Ullman can switch from radiant to ravaged in a single take. What's the connective tissue between these actors? High cheekbones, blue eyes glazed with a thousand tears unshed, and honey-colored hair that belies a lost innocence. The architecture of the Swedish face was built for film (the watery-eyed Greta Garbo was our introduction), so put a call in to Scarlett Johansson and roll camera.

WINDOW-FRAME LIGHT BOX

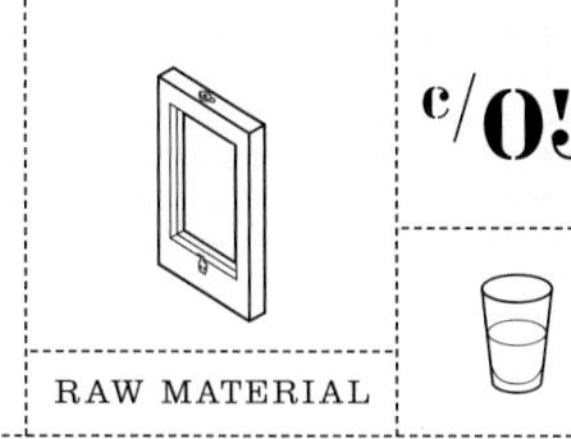

The light boxes photographers use to examine their slides make brilliant exhibitions for your own art. If only they came in custom sizes and didn't cost a mint. Enter the salvaged window, with all of its old-world charm. Enlarge a photo and print it on a transparency for a film-still effect, or illuminate your X-rays. We used off-the-shelf, prefab fluorescents (the kind that screw in under cupboards) without any diffuser on the glass so you can see the bulbs shining through. (A sheet of contact paper will obscure the glare.) Hang the heavy frame on an anchored nail and gaze at your very own window display.

WINDOW-FRAME LIGHT BOX

$47

INGREDIENTS

+ Window (ours was 18" by 22 1/2")
+ Frosted contact paper the size of window glass
+ 1/4" by 14" by 22" sheet of plywood
+ 2 1" by 3" by 14" pieces of wood
+ 2 8" fluorescent overhead light fixtures
+ 6 3 1/2" small-head nails
+ 4 1/2" small-head nails
+ D-ring with screw
+ Transparency to display

TOOLS

+ Clean rag
+ Window cleaner
+ Tape measure
+ Utility knife
+ Power drill
+ Phillips-head bit
+ Hammer
+ Double-sided tape

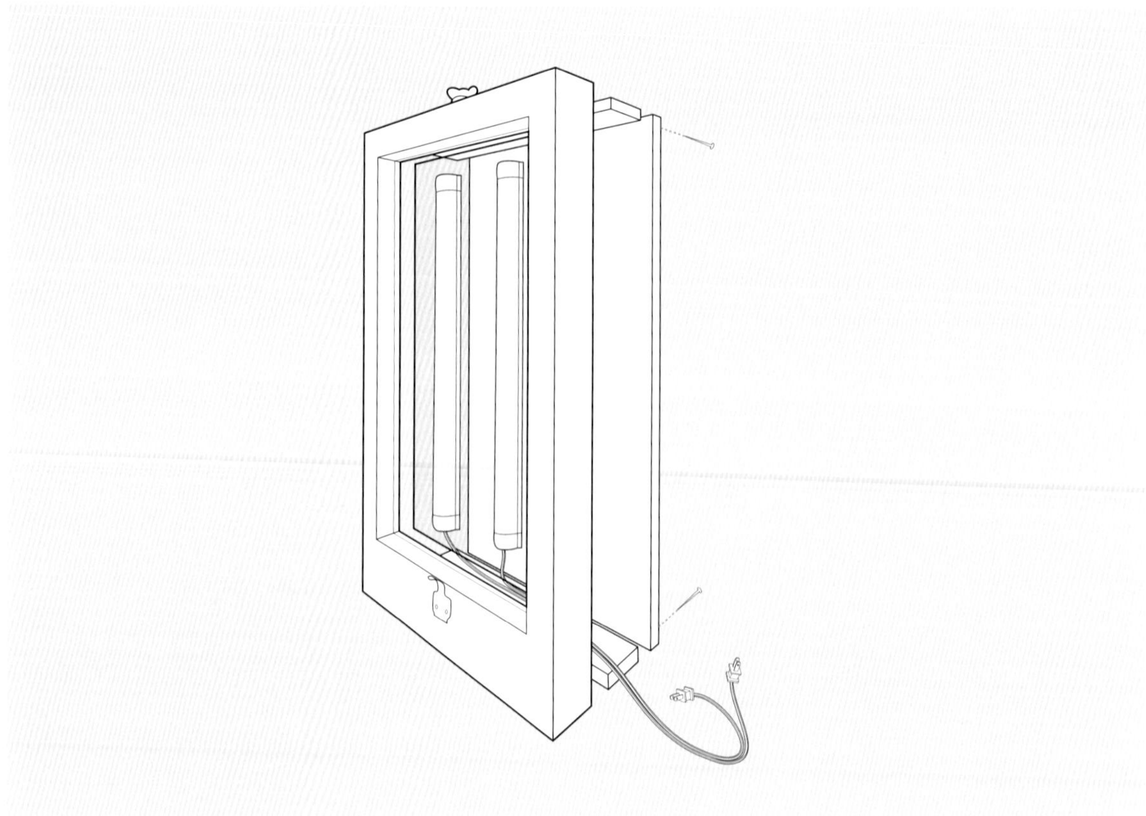

1 Clean your window.

2 Apply the frosted contact paper to the backside of the window glass.

3 Trim to size with the utility knife.

4 Mount your fluorescent light fixtures to the plywood sheet using your power drill with the Phillips-head bit. Make sure you mount them in a place that will set the tubes evenly in the frame of the window. If you don't, you won't make the tight clearance between the window and where the plywood is going to be mounted.

5 Hammer the 3 1/2" nails in the pieces of 1" by 3" wood vertically across the bottom and top of your window.

6 Flip over your plywood sheet, light fixture–side down toward the glass, and position and hammer in the 1/2" nails.

7 Screw in the D-ring centered on the backside top of the box.

8 Tape a transparency to the front of the light box, or leave bare.

9 Hang, plug in, and admire.

Lowest temperature that some glass compositions can melt: **900°F**

Percentage of glass food and beverage containers that can be recycled: **100**

WINDSHIELD FIRE SCREEN

Automobile windows are strong and tempered, and can often be found at the salvage yard. We nabbed a medium-sized back window from a vintage '85 BMW, thinking it would look great propped in front of the hearth. Alas, drilling through the thing pretty much killed its integrity as a plate of glass, and as the flames kicked up, so did the cracks. Still looks like a good backdrop for a make out session, though, no?

YES

1 We wanted to try out a simple, domestic use for a salvaged sheet of glass.

2 It's a great alternative to the very pricey (think $1000) glass fire screens on the market.

NO

1 Drilling through glass—hard to do without a proper bit.

2 Any dramatic change in temperature will make the tiniest crack grow.

DUCHAMP'S CORNER

15 Alternate Uses for Glass Jars

- Under-cabinet mounted spice rack
- Bathroom cotton-ball holder
- Sink-top soap dispenser
- Tool shop screw organizer
- Paint preserver
- Instant picture frame
- Salt-and-pepper shaker
- Pen holder
- Sand sculpture
- Homemade xylophone
- Penny collector
- Ant farm
- Flower vase
- Drinking glass
- Salad-dressing mixer

LOOKING GLASS

Salvage resources

Used Auto Glass: *(www.aachenauto.com/auto_glass.htm)*
Get your paned house windows, windshields, bottles, jars, and more through this clearinghouse.

Sea Glass Outlet *(crystalrivergems.com/products/glass/beachseaglass)*
When a whisky bottle spends centuries being tumbled by ocean waves, it ends up looking like these gems.

Percentage of recycled glass in newly purchased glass food jar: 35

WANTED:

Concrete projects*

*For Volume Two.
Please send ideas to:
ReadyMade Magazine
attn: Concrete Ideas
2706 Eighth Street
Berkeley, CA 94710

MARTINI BIRD BUFFET

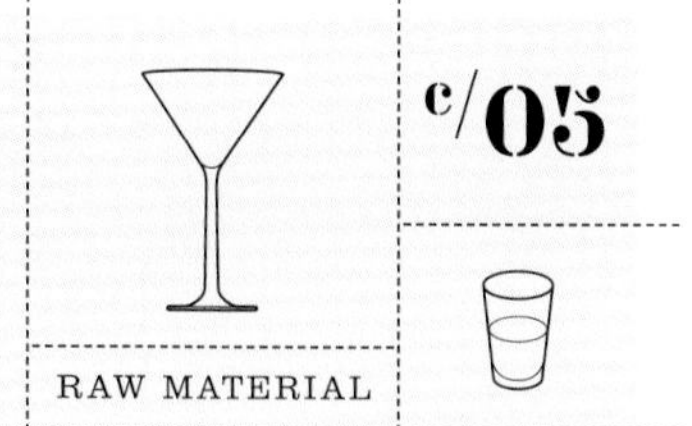

Got some scratched, chipped, or otherwise unclassy wineglasses you're looking to get rid of? Make a fancy feast for feathered friends by gluing a few stems together and hanging them from a nearby branch. The shape was inspired by the bubbly champagne waterfalls you see at gala events, but our version spills over with seed instead. Many hummingbird feeders are made of glass, so don't fear rigging your crystal outdoors. (When set close together, they double as wind chimes!) The birds will flock, eat themselves into a full-bellied stupor, and sing like drunken sailors.

Number of hours recycling one glass jar can keep a 100-watt lightbulb on: 1

MARTINI BIRD BUFFET

$8

INGREDIENTS

+ 3-5 martini or dessert glasses (a variety is better)
+ Spool of monofilament twine
+ Birdseed variety
+ Nail

TOOLS

+ Glass adhesive (glass to glass)
+ Scissors
+ Hammer

1 Clean your glasses.

2 Arrange and stack as you desire. (We recommend smaller glasses on top of larger glasses.)

3 Glue the glasses together.

4 Let glue dry overnight.

5 Cut monofilament twine to size depending on the size of your tree.

6 Loop hook around the stem of the top glass.

7 Fill tiers with birdseed variety.

8 Hang on a tree branch and wait for them to swoop down.

Percentage of glass in average household rubbish: 8

this is not a project

c/05 GLASS

THE SELF-ASSURANCE EXERCISES USED BY TOP MOTIVATIONAL GURUS

USE PENCIL ONLY

NO 2

HOW TO BREAK THROUGH YOUR OWN GLASS CEILING

THIS CAREER CATCHPHRASE has achieved such currency that it's earned its place in the dictionary. It was first floated by the equal-rights movement in the 1980s to describe the venture-no-further barrier of discriminatory practices that women and minorities rising through company ranks were subject to. Nearly twenty years later, the problem has by no means ceased to exist, but as members of the slacker generation, we're equally concerned with breaking through our own, self-imposed barriers. For us it feels more like a glass floor (tread lightly or better yet, don't move at all). If you sense that the negative voices in your head are keeping you down, want to stop living your life the way you were programmed to, have a feeling that you're responding to labeling, stereotypes, or fixed beliefs that you yourself have created, and are convinced that envious coworkers are trying to thwart your progress up the corporate food-chain—we have a plan for action! But first, take this simple test to see just how bad off you really are. (Thanks, Dr. Phil!)

(TURN OVER FOR TEST) *****Please do not fold or mutilate*

DO NOT WRITE IN THIS AREA

FEED THIS DIRECTION

DO NOT WRITE IN THIS AREA

DO NOT WRITE IN THIS AREA

STEP ONE

Circle all the words you think describe your potential.

pretty / attractive / beautiful / cute / nice hair / bad hair / cool / spiritual / friendly / faithful / slut / leader / strong / supportive / honest / decent / warm / loving / tender / good crier / caring / good cook / cordial / welcoming / cheerful / passionate / fiery / enthusiastic / arrogant / egocentric / martyr / pity-worthy / humane / selfless / philanthropic / smart / dependent / accomplished / free / gentle / abrasive / thoughtful / submissive / domineering / creative / compassionate / self-sufficient / private / conventional / rocket scientist / dashing / clever / quick / sneaky / crafty / charming / tidy / paranoid / obsessive / compulsive / binger / alert / reliable / inspired / inventive / resourceful / ingenious / productive / exciting / energetic / lively / vigorous / bouncy / ecstatic / cheery / sane / rational / sensible / normal / filthy / rich / wealthy / prosperous / full / gorgeous / abundant / fruitful / chubby / deep / prolific / useful / helpful / constructive / functional / worthwhile / complete / capable / inspiring / proud / approachable / at peace / giving / nurturing / whole / perfect

COUNT THE NUMBER OF WORDS YOU CIRCLED IN STEP ONE. **THIS IS YOUR POTENTIAL SCORE =**

STEP TWO

Now circle the words that describe how you actually are today.

pretty / attractive / beautiful / cute / nice hair / bad hair / cool / spiritual / friendly / faithful / slut / leader / strong / supportive / honest / decent / warm / loving / tender / good crier / caring / good cook / cordial / welcoming / cheerful / passionate / fiery / enthusiastic / arrogant / egocentric / martyr / pity-worthy / humane / selfless / philanthropic / smart / dependent / accomplished / free / gentle / abrasive / thoughtful / submissive / domineering / creative / compassionate / self-sufficient / private / conventional / rocket scientist / dashing / clever / quick / sneaky / crafty / charming / tidy / paranoid / obsessive / compulsive / binger / alert / reliable / inspired / inventive / resourceful / ingenious / productive / exciting / energetic / lively / vigorous / bouncy / ecstatic / cheery / sane / rational / sensible / normal / filthy / rich / wealthy / prosperous / full / gorgeous / abundant / fruitful / chubby / deep / prolific / useful / helpful / constructive / functional / worthwhile / complete / capable / inspiring / proud / approachable / at peace / giving / nurturing / whole / perfect

COUNT THE NUMBER OF WORDS YOU CIRCLED IN STEP TWO. **THIS IS YOUR ACTUAL SELF SCORE =**

FEED THIS DIRECTION

DO NOT WRITE IN THIS AREA

STEP THREE

The direness of your state of affairs is determined by the percentage of words you circled in Step 2 as compared to the total words scored in Step 1.

The calculation: The number of words circled in Step 1, minus the words circled in Step 2, divided by the total number of words equals your percentage. If you aren't comfortable doing the math yourself, ask someone who's clearly smarter and more accomplished than you.

$$\frac{\text{POTENTIAL SCORE} - \text{ACTUAL SELF SCORE}}{\text{POTENTIAL SCORE} + \text{ACTUAL SELF SCORE}} = \quad \%$$

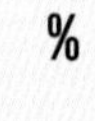

THE TALLY

If your score is:

90–100%
You are brimming with confidence and self-worth. Go forth and conquer.

75–89%
You are in fairly good shape, consistently egging yourself on and succeeding most of the time.

50–74%
You have realized some good aspects of who you are. However, you are plagued with enough self-doubt to sabotage yourself from time to time.

35–49%
You are limiting yourself and are guided by a distorted sense of who you actually are. Much work is needed.

1–34%
You are living in your fictional self. You are clearly damaged and wasting precious life energy. Your power is infected with fiction.

For those in the red (1 to 50 percent), here are a few pointers on how to reach your full potential as a peace negotiator, world-class athlete, Pulitzer Prize winner and Top 10 seller on Amazon.com, sexiest person alive according to *People*, wildly successful entrepreneur, Churchillian orator, strategic adviser to world leaders, and TV-authority-cum-commentator on all subjects:

1. Think differently.
Most people are stymied by perceptions of themselves that have become fixed in stone. These can easily be changed. Think you are shy? Tell yourself you're really outgoing at least ten times a day, force yourself to speak up at the moment you think of something to say, and be the first to say hello at least five times a day. You'll be an extrovert before you know it.

2. Once you're succeeding, protect yourself from saboteurs.
Are friends, family, and coworkers behaving as though your new, more fully realized persona is not very appealing? Are they asking if you've gone on meds or made a "lifestyle change" they should know about? Do they fear that you'll only be "hurting yourself"? Envy is the plate of cold food that gets served up as soon as you start pushing ahead of others. They will be threatened by your shiny new self-worth because they are still mired in their own swamp of negativity. They want you to remain below them, or at least on their level, and if you don't comply, they will attempt to destroy you. It's difficult to beat back this kind of resistance, but when you encounter it, pretend that it's the work of aliens who have come to eradicate humankind and take over the Earth. It is your job to detect the aliens and subvert their alien ways. If you cave, the entire human race will be lost!

LOOK AT THAT: YOU ARE A SUPERHERO!

DO NOT WRITE IN THIS AREA

FEED THIS DIRECTION

DO NOT WRITE IN THIS AREA

 c/05

Why Don't I Do This Every Day?

TESTIMONIAL # 0085

NAME: **Pete Glover**

FROM: **Pittsburgh, PA**

OCCUPATION: **Self-Proclaimed Hustler**

Pete, why don't you do these things every day?

In today's environment of multitasking, who's got the time? I'd love to try my hand at these projects, but between relaxing, kicking back, cooling out, slacking off, taking it easy, lounging, lurking, and loitering, I barely have time to keep it real. And the Internet isn't going to just surf itself, gnome sane, bro?

1

MEASURING JAR

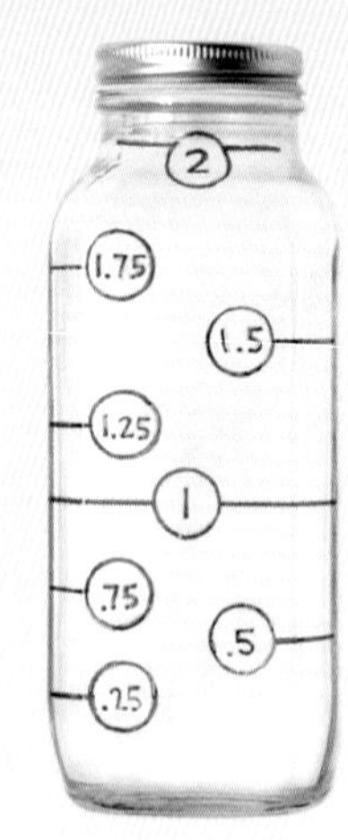

INGREDIENTS

+ Empty jar

TOOLS

+ Sharpie
+ Rubber band
+ Measuring cup

01 Clean and dry your jar.

02 Measure and pour increments of 1/4 cup water into your jar with your measuring cup.

03 Wrap the rubber band around the jar. This is your guide for drawing straight lines around your jar's perimeter.

04 With each pour, move your rubber band just under the height of the water and mark with your Sharpie. Do this until your jar spills over. (For permanent and dishwasher-safe markings, use Pebeo Vitrea 160 paint markers found at *www.dickblick.com*.)

05 Use your new measuring jar to bake a cake.

2

COLORED GLASS PLATES

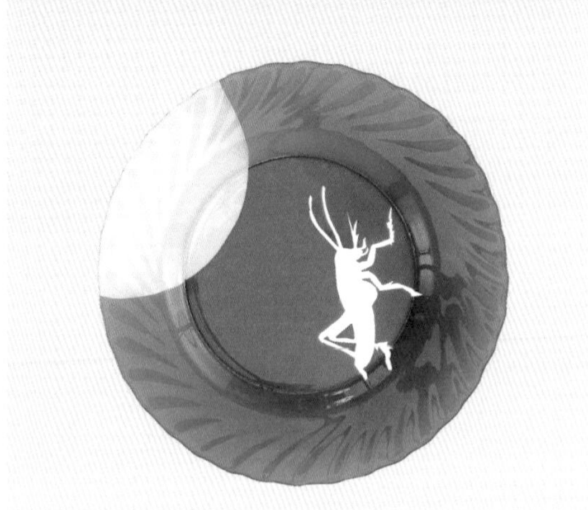

INGREDIENTS

+ Assorted glass plates
+ Clear contact paper
+ Frosted-glass spray paint

TOOLS

+ X-Acto knife
+ Newspaper

01 Clean the plates.

02 Adhere the clear contact paper to the backside of a plate.

03 Cut away the areas you want color on the plate.

04 Place the plate contact-paper-side-up onto the newspaper, outside, on a flat surface.

05 Spray several light coats of the paint.

06 Dry overnight for best results.

07 Peel off the contact paper.

08 Flip over and display or serve.

We really should, you know.

3

JUG STEPPING-STONES

INGREDIENTS

+ 4 large-diameter juice or pickle jugs with caps
+ 3 smaller-diameter juice jugs with caps
+ Images cut to the shape and size of the bottom diameter of the jars
+ Packing peanuts
+ Few sheets of cardstock

TOOLS

+ Shovel
+ X-Acto knife

01 Empty, rinse, and dry the jugs and caps.

02 Insert your cardstock image, cut to the size of the bottom of a jar with the image facing the glass.

03 Pour packing peanuts on top of the paper insert to hold paper in place. Fill to the top. Put the cap back on.

04 Dig a hole in the ground where you want the stepping-stone to live.

05 Bury the jug upside down, ensuring the bottom surface is at or slightly above ground level.

06 Repeat steps 2 through 5 for the other jugs, and arrange in the ground as you wish.

Tarry no longer, chop chop!

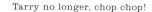

Chapter: 06

FABRIC

READY-TO-WEAR

After Eve trades in her innocence for knowledge, she and Adam suddenly realize they need new outfits. They grab the biggest leaves they can find in Eden (fig variety) and sew them into loincloths. Observing their shame and their feeble attempt to cover up, God provides them with leather duds (raw animal skins) that are less revealing and won't wilt. The Provider wants to be sure they know that only He can fashion wraps to cover their sin, and that it comes at a price—the suffering of other creatures.

After the couple goes forth and multiplies, there's suddenly demand for a lot of animal skins, and everyone and their brother's five wives wants to be a shepherd. Abraham, Rachel, Jacob, Moses, and David all do it. And so it comes to pass that wool is a major commodity in Israel. Mesha, king of Moab and a great sheep breeder, is under pains to deliver "the wool of a hundred thousand rams" to the king of Israel annually. Linen, which was spun from the flax plant, was also popular for head wraps, togas, and the like. When the high priest Aaron entered a holy place, he put on a linen coat and girdle (the first tallith) and a linen cap (the first yarmulke).

Cotton and silk have been farmed for millennia as well. But our now ubiquitous synthetics—Spandex, Lycra, nylon, Gore-Tex, and all varieties of polyester—made their debut only in the last century. We love all of our new breathable, stretchy, waterproof, wrinkle-free clothing—there's just too much of it! The average American throws away sixty-eight pounds of garments and rags each year. Textile recycling, which is very little known but occurs in most cities, removes 2.5 billion pounds of consumer clothing waste headed for landfills—ten pounds per American. Pretty anemic when you consider that around 145 billion pounds of recyclable clothing are still sent to the dump every year.

So, what happens to your Michigan Wolverines T-shirt that never found a new owner after you donated it to the Salvation Army? Some gets bought up by used-clothing dealers and fiber recyclers—most of them small, family-owned businesses. They purchase our old shirts as "mixed rags" surplus or work out agreements with city recycling centers to set out collection bins in places where people might donate. About 20 percent of the clothes they purchase are turned into cleaning rags sold to industry for mechanics and janitors. The rest of it—almost half—is exported to the world's poorest countries. So your Wolverines tee could very well now be worn by someone in India.

Threads that are too damaged or stained to be of any use can be pulverized into shoddy, a fine fiber used in mattresses and insulation. Industrial farmers use shoddy as roughage for cattle. And it makes plastics and paper more durable when mixed into a recycling batch. So fill those garbage sacks with your tired get-ups and get thee down to thine local charity. They'll gladly take last year's black off your hands. As for the rest—old blankets, towels, and tablecloths—make one of the projects you see here.

Around 145 billion pounds of recyclable clothing are still sent to landfills every year.

Photo: David Byrne

A BRIEF HISTORY OF FABRIC

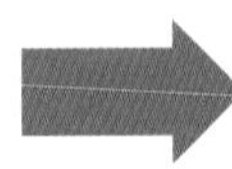

5000–4000
Wool was the first animal fiber turned into cloth. It was spun into yarn by early shepherds, who were the first to domesticate and breed sheep for their fleece. Once land cultivation began, one of the most common crops, flax, was turned into the earliest vegetable fiber to be woven.

2800
Chinese cultivate silkworms, feeding them on mulberry leaves, unwinding their cocoons to produce long strands of silk fiber. In the Roman empire, silk is sold for its weight in gold.

AD 50
First wool factory is built by the Romans, who needed warm woolens if they were going to go off and build an empire in foreign climes.

1700s
English developed machines to spin thread and weave cloth in large quantities.

1850
Levi Strauss, a twenty-year-old Bavarian immigrant, leaves New York for San Francisco with a supply of rough canvas to sell to gold rushers who need tents and wagon covers. When he arrives, the prospectors tell him, "You should have brought pants!" Strauss had his stores of canvas fashioned into pants, but miners complained that they chafed, so he substituted a twilled cotton cloth from France called "*serge de Nimes*," which became known as denim. The red tab attached to the left rear pocket was created in 1936 as a means of identifying Levi's jeans at a distance.

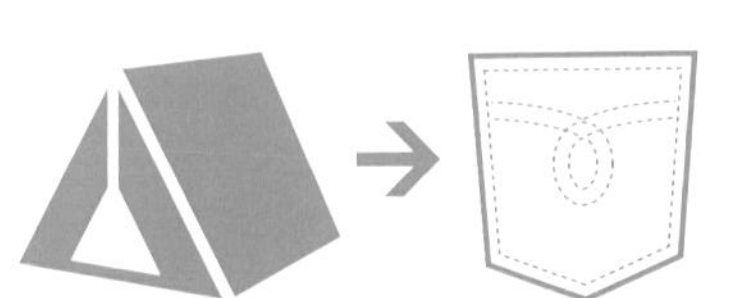

1953
First commercial polyester fiber production in the United States by the DuPont Company.

What's the connective thread between

shearing sheep and the space suit?

Here's how it went:

12,000 BC

Some authorities claim that it was likely that the Egyptians had cotton as early as Abraham, and evidence of cotton has been found in Mexican caves (cotton cloth and fragments of fiber interwoven with feathers and fur), which dates back to approximately seven thousand years ago. There is clear archaeological evidence that people in South America and India domesticated different species of cotton independently.

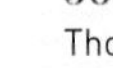

300s

Though cotton was cultivated back in 5000, it wasn't until now that Alexander the Great's army brought cotton goods (no doubt forcibly) back to Europe. Only the rich could afford them, natch.

1823

Scottish chemist Charles Macintosh patents a method for making waterproof garments by using rubber dissolved in coal-tar naphtha for cementing two pieces of cloth together.

1823

In Farmington, Maine, fifteen-year-old Chester Greenwood, a grade-school dropout, invents earmuffs. Farmington is now considered the earmuff capital of the world, and there's an annual parade on the first Saturday in December to mark Greenwood's accomplishment.

1930s

Synthetic fiber nylon is introduced as an early substitute for silk and soon becomes the choice material for stockings. Nylon is the second most commonly used man-made fiber in the United States, after polyester.

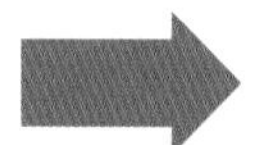

1989

WL Gore & Associates introduces a membrane technology engineered to be a breathable water-, and wind-proof material. They register the trademark as GORE-TEX, and the sporting-apparel industry goes nuts for the stuff.

PRESENT DAY

The fabrics just keep getting smarter. Some recent introductions include bug-repellent and UV-protected weaves, luminescence, electronically interactive "smart" fabrics, and textiles that clean themselves. Now that's what we're talking about.

Pr
Pc
Ml
Wd
Gs
Fc
CHEMICAL
BREAKDOWN
FABRIC

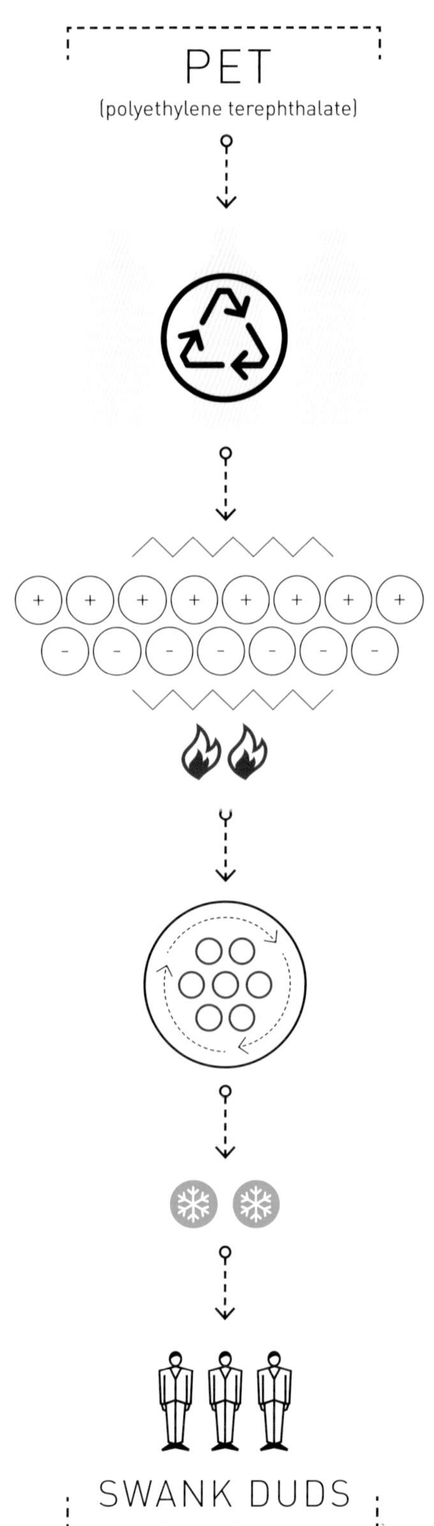
PET
(polyethylene terephthalate)
SWANK DUDS

MAKING POLYESTER

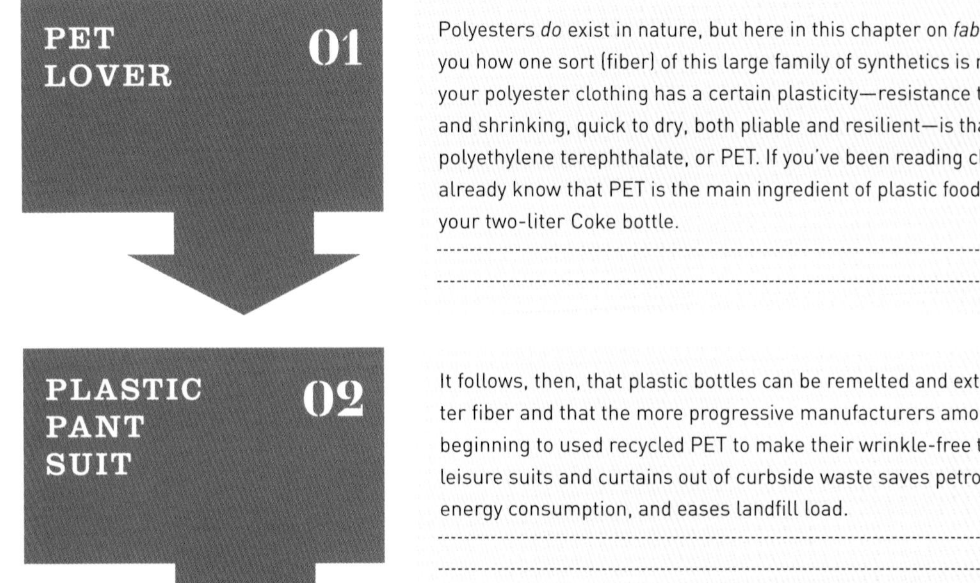

PET LOVER 01

Polyesters *do* exist in nature, but here in this chapter on *fabrics*, we will show you how one sort (fiber) of this large family of synthetics is made. The reason your polyester clothing has a certain plasticity—resistance to stretching and shrinking, quick to dry, both pliable and resilient—is that it's made from polyethylene terephthalate, or PET. If you've been reading closely, you already know that PET is the main ingredient of plastic food packaging, like your two-liter Coke bottle.

PLASTIC PANT SUIT 02

It follows, then, that plastic bottles can be remelted and extruded as polyester fiber and that the more progressive manufacturers among us are beginning to used recycled PET to make their wrinkle-free threads. Making leisure suits and curtains out of curbside waste saves petroleum, reduces energy consumption, and eases landfill load.

CALLING ALL CRUDE OIL 03

Not just for gas pumps anymore! Petroleum goes through an esterification process to become PET (when you break it down, the name spells it out: *Poly* means many, and qualifies the linkage of several esters within the fiber). Ester groups in the polyester chain are polarized with negative charges and positive charges that are drawn to each other, bonding in a chemical reaction that forms crystal-like chains, which are melted down, purified, transesterized, and polymerized (this is serious chemical engineering business) before being sent to the spinner.

THE MAGIC OF MELT SPINNING 04

The fiber-forming substance is melted for extrusion through the spinneret and then directly solidified by cooling. Melt spinning can churn out different shapes that are used for carpet and other textiles (round, trilobal, pentagonal, octagonal). Trilobal-shaped fibers, for example, reflect light and give the yarn a certain sparkle.

LAMP COZY

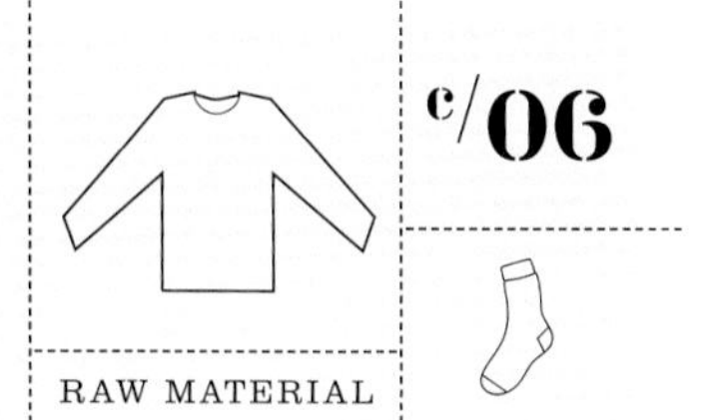

From mid-century designer and sculptor Isamu Noguchi's original to the ubiquitous Ikea knock-off to our *ReadyMade* update, this cocoonlike lampshade has had many lives. When you want a softer, woollier glow than plain white paper can provide, sew together the sleeves of old sweaters into a stretchy tube and slip it over your frame. The knit diffuses glare and adds a warm, multilayered glow. Use a looser weave in a few sections, as shown here, to allow more light to filter through. Fuzzy logic at its best.

Number of animals required to make one cashmere sweater: 4–6

LAMP COZY

INGREDIENTS

+ Floor lamp
+ 3 old sweaters

NOTE: We used a floor lamp from Ikea and bogarted the wire from the frame. You can adapt this design using wire hangers or tomato wire if you want a more challenging project.

TOOLS

+ Scissors
+ Sewing machine
+ Thread

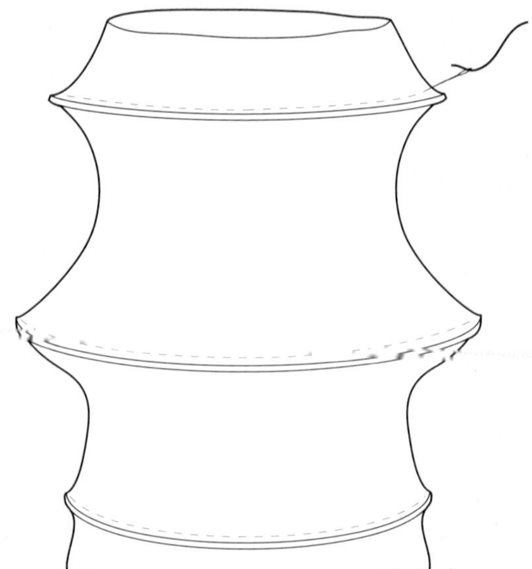

FIGURE A

1 Remove the shade and strip it until you have a naked frame. (We pulled off the paper shade from the metal wire rings.)

2 Lay out the sweaters you'll be using for the shade, and arrange in your desired color orientation.

3 Test out the tautness and opacity of sweater material when the sleeves are stretched across your rings. You want the knit to be taut enough to hold the ring stationary without your assistance. Depending on the size of the sweater, the sleeves will probably work best for everything but the widest part of the lamp.

4 Cut off the parts of the sweaters you want to use. Starting from the top of the lamp and working your way down, sew the knit cylinders onto the wire rings (Figure A).

What a sweater made of angora rabbit hair must have to be legal: A label saying as much

FIGURE B

FIGURE C

5 For a clean look when you add the next sweater, make sure the raw-cut seams are on the interior of your shade. This is achieved by working inside out, in an orderly fashion (Figures B, C). Vary the width of the rings and the color and texture of the knits for the best effect as you work your way down. Also vary the distance of knit between the rings.

6 Measure the approximate length you have sewn together after you have finished three or four rings to anticipate how many segments you have left to create. (Our shade's bottom rung was tension-set to the lamp stem, which gave us some freedom with the length.) Finish sewing on the remaining rings.

7 Plug in, turn on, and cozy up with a good book.

Percentage more that Spandex will stretch over the same amount of rubber: **500**

DENIM DOG BED

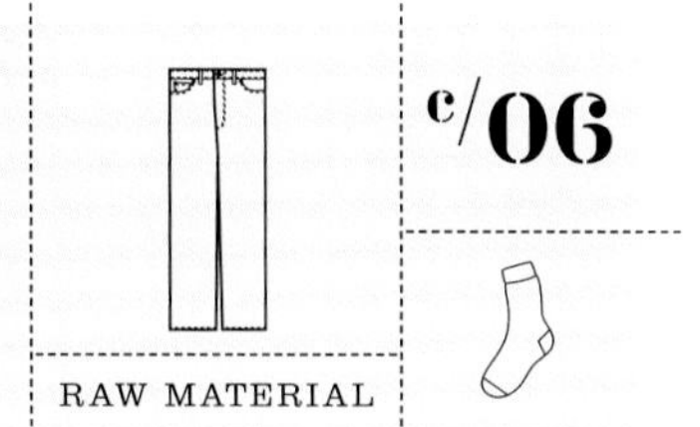

Denim was originally conceived for miners and farmhands who needed their clothes to stand up to the hard labor of digging, herding, and rough rides in the saddle. That tough-skin quality is ideal when making furniture for Fido. Sew together the legs of worn-out jeans as bolster walls (your furry friend will attempt more rippage once it's in his corner, anyway) and the main bed from pants seats, then stuff the whole thing with old rags or T-shirts. This version makes a lounge for a midsized hound, like a boxer or springer spaniel, so adjust as necessary for your pup. Best friends forever!

Number of days the cotton plant requires from planting to full maturity for harvest: 180–200

DENIM DOG BED

INGREDIENTS

+ 2 pairs of jeans (ours were size 44)
+ 2 16-ounce bags of synthetic fiberfill

TOOLS

+ Scissors
+ Pins
+ Sewing machine
+ Thread
+ Blue masking tape

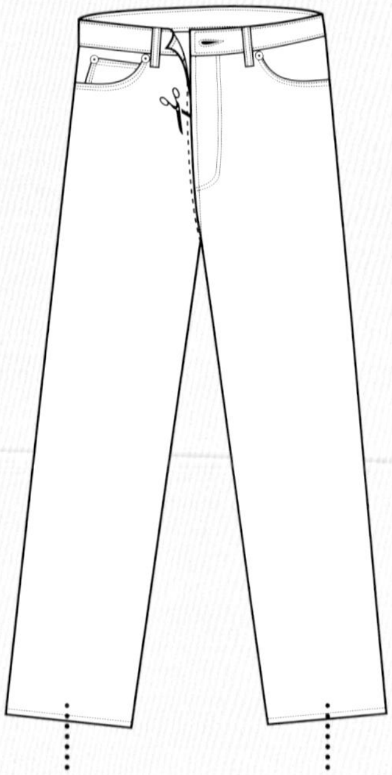

TOP SIDE OF BED WALL

FLY SIDE

FIGURE A

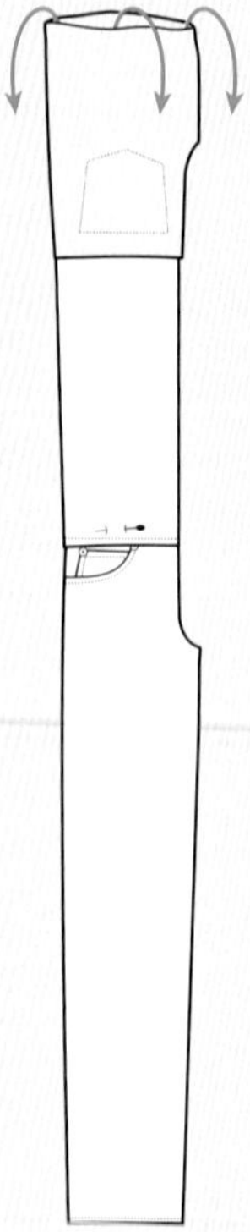

FIGURE B

1 Taking care to cut as close to the double-sewn outer seams as possible, cut open one pair of jeans down the crotch and up the backside (Figure A).

2 Cut away the zipper fly and waistband to keep the denim easy to work with.

3 Lay out the two split legs, head to toe, and fly side down (Figure B). Connect the legs. Pin together the legs so they won't rotate out of place as you sew one side together.

4 Peel one leg over the other so the front sides of both legs are facing each other to sew a hidden seam (Figure C).

5 Sew the other two ends together to form the doughnut shape of your pet bed wall. Leave a fist-sized opening for the filling.

6 Cut open the second pair of jeans on the inner inseam, making sure to leave the double-layered seam on opposing sides. This will be the floor of the bed (Figure D).

7 Splay open and sew opposite legs together down the middle, making a wide, flat piece that will serve as the floor of the bed. We elected to have the inside of the jeans serve as the bed

Thinking it would make them stay put, she told her two sons that the rug in their bedroom would levitate if they sat on it quietly and concentrated hard. The next morning, she found them lying flat on the rug, fast asleep, with arms outstretched in flying formation. Their pajama pants were soaked through.

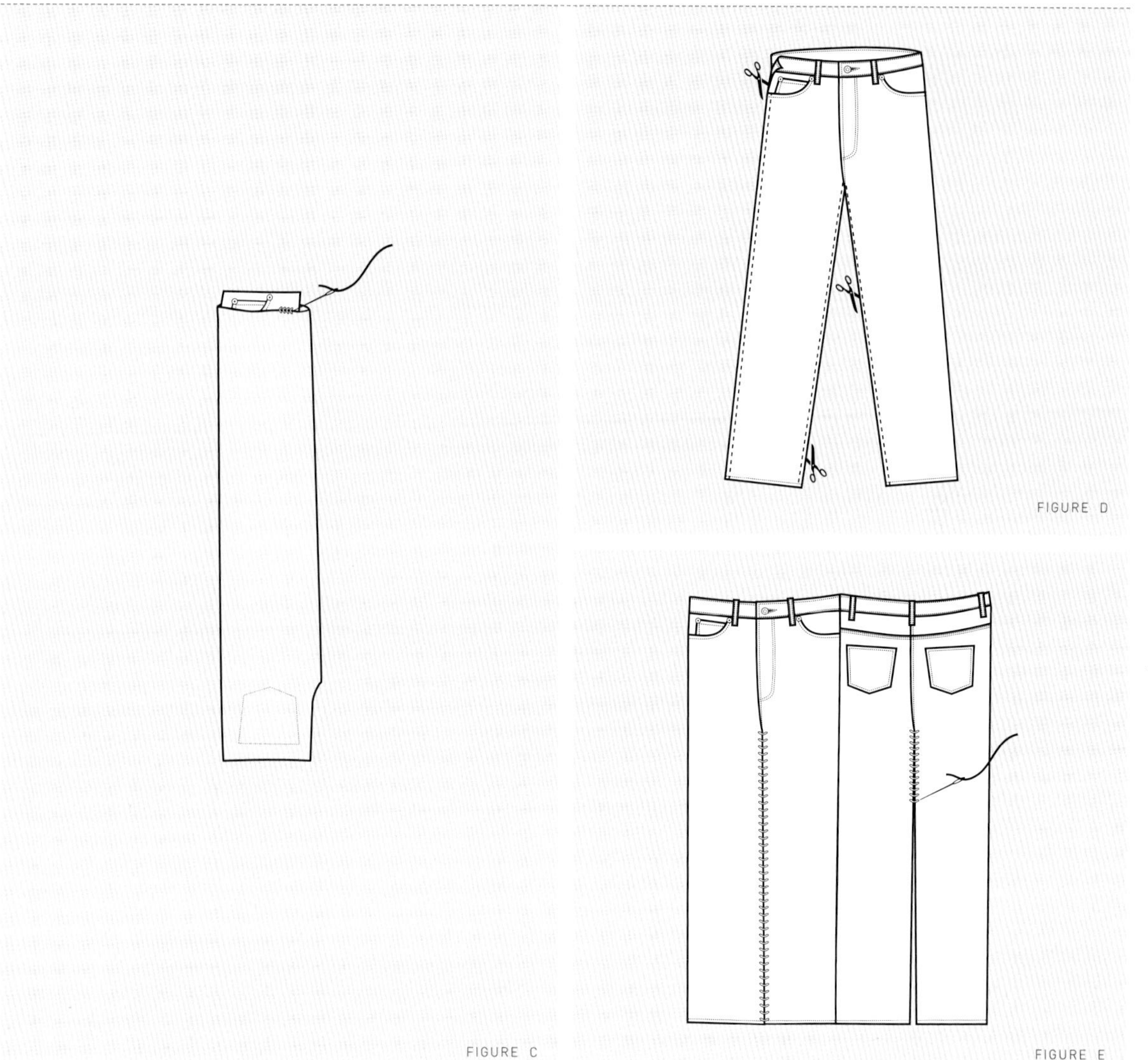

FIGURE C

FIGURE D

FIGURE E

floor for contrast (Figure E).

8 Fold in half and sew the bottom hems together.

9 Loosely arrange the doughnut-shaped wall on top of the fabric floor and place blue masking tape on the floor piece around the shape of your arranged bolster.

10 Remove bolster and sew the floor shut on the inside of the blue tape, leaving a fist-sized opening for the filling.

11 Cut away the excess fabric. You now have the floor of the bed (Figure C).

12 Pin the bolster wall, crotch side down, in place to the floor near the middle outer edge of the bolster. Once the walls are filled, the floor seams will vanish underneath the bed.

13 Sew the floor to the bolster and remove the pins.

14 Stuff the walls and the floor to desired firmness.

15 Machine-sew the opening shut.

16 Call your pooch over to sit in his new lap of luxury.

Percentage of South Carolina's cotton production lost between 1920 and 1922 due to the boll weevil: **70**

this is not a project

c/06 FABRIC

LEARN FROM OUR GREATEST CHARLATANS AND PUT YOUR BLUFFING SKILLS TO GOOD USE

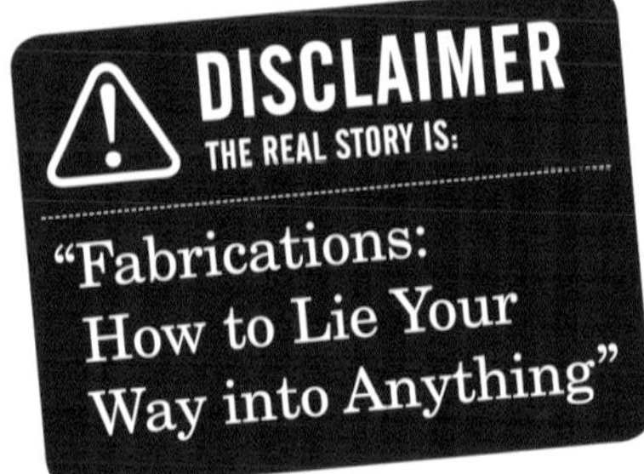
DISCLAIMER
THE REAL STORY IS:
"Fabrications: How to Lie Your Way into Anything"

Nobody ever got anywhere in life without fudging the truth. It's one of our most underrated skills. And there's a full bouquet of deceit, from the out-and-out lie to the wilting half-truth to the little fictions that add color to our stories. We start your lesson plan off with a few case studies of bluffers from recent history who were famously busted. We then offer some how-tos that will apply in common life situations, as well as a few tips on how to keep your poker face so you don't meet the same end they did. Choose whatever degree of falsification suits you, and please play nice. Do not use your newfound talent to cheat or steal or hurt people. And if you do, don't come shaking your subpoena at us.

LIES AND THE LYING LIARS WHO TELL THEM

Lying for love and money:

Mata Hari

Magaretha Zelle, a Dutch woman living in the East Indies, divorced her husband, moved to Paris, rechristened herself Mata Hari (literally, "eyes of dawn" in Malay), and began a riotously popular stint as an exotic dancer. Her stage persona was an Indian-born girl of Brahman caste who lived in the temple of the god Siva. She dressed in a clingy body stocking, wore bejeweled cuffs on her arms and neck, and slunk around provocatively. But the performances were not Mata Hari's finest act. She is perhaps the most famous spy in history. She traveled with her act to Germany and took on a passel of lovers along the way, most likely becoming a mole for the German government during World War I. In Paris, she began an affair with a twentysomething Russian pilot who flew for the French. When her soldier was wounded and she requested permission from the Gallic authorities to be allowed to see him, they offered a trade—visitation rights for spying on her acquaintance the crown prince of Germany, plus a sackful of francs to sweeten the deal. No one really knows which side Mata Hari ended up undercover for, or if she was really spying at all, but she was having plenty of fun along the way, as the affairs with multinationals just kept coming. Then, en route to Germany to see the crown prince, Mata Hari was deported by the British (who had their suspicions) to Spain. Once there, she took a German military attaché as a lover, the final affair that would seal her fate. He sent a message to Berlin that the spy had proved valuable in a code he knew the Allies could decipher. Upon returning to Paris in 1917, Mata Hari was taken into custody by the French. No evidence was ever found against her—she claimed the secret ink in her room was stage makeup, and that the deutsche marks she'd received were a gift from her lover, not payment for services rendered. That didn't stop a court martial, which sentenced her to execution. Refusing a blindfold, she blew a kiss at the twelve-man firing squad before they mowed her down—a liar to the end, and proud of it.

Lying for career advancment:

Richard Nixon

A mere coincidence that Nixon's Committee to Re-elect the President (formally CRP) was referred to as CREEP? We don't think so. Through his CREEP cronies, Nixon sent a group to burgle the Democratic National Committee offices in the Watergate building, implanting microphones rigged in ChapStick (hello, MacGyver!) and nabbing files. In an effort to cover up the scandal, Nixon persisted in a series of lies that included his famous televised address, in which he pleaded, "I am not a crook." The deception was eventually brought to light by Bob Woodward and Carl Bernstein of the *Washington Post*, with help from an anonymous source code-named Deep Throat. The burglary wasn't the only dirty trick undertaken by CREEP, mainly under the aegis of Howard Hunt and G. Gordon Liddy, who had formerly worked in the White House, in the Special Investigations Unit, which looked into leaks of covert information and pursued operations to counter Democrats and antiwar protesters (most famously by trashing Daniel Ellsberg's psychiatrist's offices in search of evidence that he had furnished classified Defense Department documents to the *New York Times* detailing the government's role in engineering U.S. involvement in the Vietnam War). Members of the Hunt-Liddy unit were known as the "plumbers." (Who comes up with these names?) In January 1973, the original burglars, along with Liddy and Hunt, were convicted of conspiracy, burglary, and wiretapping. After that, fatal hemorrhaging of Nixon's administration began. He attempted to save his own skin by holding on to tapes, recorded at the White House, that revealed his involvement in the Watergate break-in, claiming executive privilege, and forced the resignations of top cabinet members. But the President's runarounds eventually led to impeachment proceedings and, soon after, his resignation in August 1974. To his death, Nixon claimed he was honest and innocent (though he accepted a presidential pardon from his successor, Gerald Ford, protecting him from prosecution).

If only poor Richard had lived in our times. With the threat of terrorism, lying while in office is now A-OK. When it comes to subterfuge, the Bush posse makes Nixon's cabinet look positively Washingtonian.

Lying to tell a story:

Stephen Glass

In the late 1990s, a twenty-five-year-old reporter at *The New Republic* was getting some great scoops. He wrote dozens of pieces for national publications in which he fabricated quotes, added dramatic twists, and made up sources to bolster his claims. He had keen instincts for what kinds of stories sell to editors, and when he couldn't find those stories, he simply made them up. But by 1998 the jig was up, and the *Republic* fired Stephen Glass for journalistic fraud. One of his best ruses included setting up a fake company Web site and voicemail account for Jukt Micronics, both of which snowed the *Republic*'s fact-checkers. The article in question was a story about a fifteen-year-old who hacked into a corporate server and extorted money. "Hack Heaven" is told in first person, as though Glass were there when the crime went down, and is full of descriptive flourish. Here's how it begins:

> *Ian Restil, a 15-year-old computer hacker who looks like an even more adolescent version of Bill Gates, is throwing a tantrum. "I want more money. I want a Miata. I want a trip to Disney World. I want X-Men comic [book] number one. I want a lifetime subscription to* Playboy, *and throw in* Penthouse. *Show me the money!"*
>
> . . .
>
> *Across the table, executives from a California software firm called Jukt Micronics are listening and trying ever so delicately to oblige. "Excuse me, sir," one of the suits says, tentatively, to the pimply teenager . . .*

Suspicious of the story, a reporter at *Forbes* investigated—and proved it to be a pack of lies. Digging a little

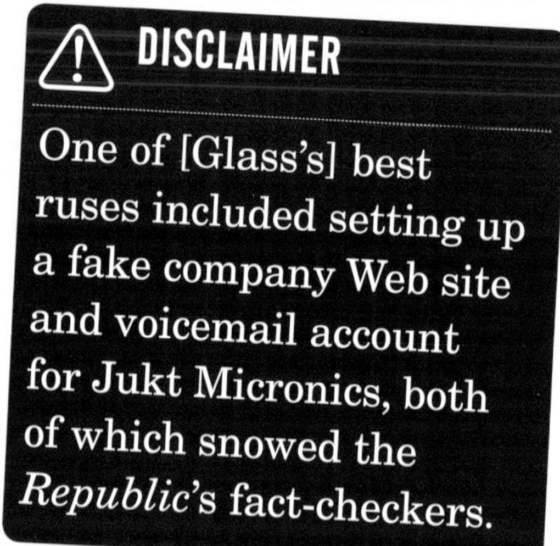

deeper, editors at *The New Republic* found that twenty-one of forty-one stories written by Glass for the magazine had been padded with fabrications.

For a consummate liar, the only way to get out of such a pickle is to come clean, then make people feel compassion by calling it a pathology and a professional hazard, something beyond the liar's control. In the end, Glass's ambition was not in the least deterred by being found out. He finished a law degree at Georgetown University (attorney-client privilege can amount to lying—he had all the right credentials), then became a minor celebrity, appearing on TV to promote *The Fabulist*, a "novelized biography," and to comment on *Shattered Glass*, a movie based on the scandal.

How to be a good bluffer

1. Fill your story with lots of details. Embellishments are easy to make up on the spur of the moment, and it's hard for us to believe that someone would expend the effort required to make something seem so probable in advance. Stephen Glass's faked accounts had the air of truth because they included so many apparent facts: company and source names, scene-setting details, descriptions of how people looked down to the brands they wore. If you don't have that, any follow-up questions will have you stumbling for answers. Your composure will crack like a plastic knife trying to cut through steak.

2. Recite your lie so often that you begin to believe it yourself. As was the case with Nixon, you really have to brainwash yourself to be a successful fibber. The human brain wants to tell the truth. It requires an enormous outlay of energy to counter that drive. Successfully convince yourself of your own perjuries, and you're bound to convince others, too.

3. Always look your dupe in the eye. The eyes, as Shakespeare would have it, are windows—they reveal our interior lives. It's tough to gaze directly at someone, even the people closest to you. (As animals we're not far from dogs, who perceive an unwavering look as a threat.) So we'd expect those who are anxious about being found out to be shifty-eyed, lacking confidence. Try concentrating on something about your dupe's eyes as you lie—the color, the rings, crow's feet—and you're sure to be believed.

4. Pretend you're playing a game. This is advice we once received for how to relax when you're on TV. Imagine that you're having a bit of fun with friends and that the object of the fun is to make up a great story, and you'll be surprised how easily you become a well-regarded pundit.

5. Cry. When all else fails, come unhinged. If they're still casting doubt, take a pregnant pause, think about something really sad (your childhood dog, old men), and let the floodgates open. It takes a real bully to doubt someone showing signs of severe emotional distress.

Sometimes lying is the kinder gentler way: The road to hell is paved with good-intentioned truth-tellers. Sometimes a little white lie protects people from a big black depression. Good occassions for lying include: when someone delivers a bad performance onstage; when answering the question "Do I look fat in this?"; when you want to get into an event but are not on the list; when you are somewhat but not wholly qualified for a job; when you're planning a surprise party; when you're having an affair with someone at work; when you're the one who stepped in it; when a child asks why he or she can't have something and there is no reason that will ever make sense to a child.

DUCHAMP'S CORNER

16 Alternate Uses for Socks

Eyeglass case
Leg warmer
Pencil holder
Holiday stocking
Elbow patch
Cell-phone cozy
Change saver
Hand puppet
Hair mop
Wine diaper
Wrist cuff
Jewelry hider
Headband
Shoe shiner
Stain applier
Soap saver

CLOTHES HORSE

How to recycle your threads

Good Will Industries *(www.goodwill.org)* The old gray lady of thrifty sweaters, shirts, jeans, and other textiles. Find a location near you at their comprehensive site, and give a little while you're at it.

Clothing Recycling Bins *(www.recyclingbin.com/clothing_recycler1.htm)*

LACE FRUIT BOWL

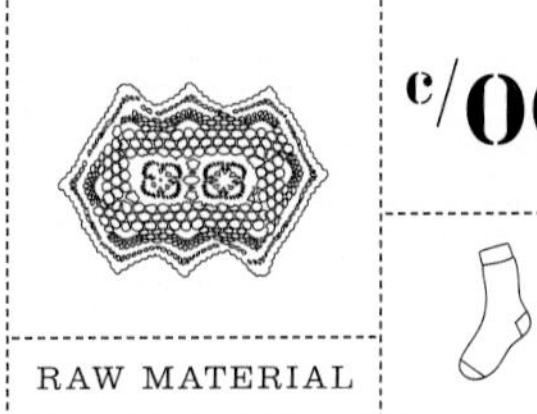

The era of doilies and lace tablecloths may be over, but don't let us catch you tossing out Grandma's tea-stained table coverings. Instead, make them the basis for a science project. Lay a cut-out square of tatting in a cardboard produce holder or other pleasingly shaped mold, don your safety gloves, and pour on the epoxy resin. Inspired by the open netting of the Knotted Chair by Droog Design frontman Marcel Wanders, we created this crocheted fruit bowl. Airflow through the mesh increases the life span of the produce, and you can stack your apples into a perfect pyramid.

Number of times that linen is stronger than cotton: 2–3

LACE FRUIT BOWL

$16

INGREDIENTS

- 20" by 11" lace remnant
- 4 plastic kitchen trash bags
- 2 cardboard fruit trays from your local market
- Epoxy compound (found at specialty art suppliers)

TOOLS

- Scissors
- Safety gloves
- Disposable mixing dish
- Disposable plastic utensil for stirring
- Disposable paintbrush
- Shears

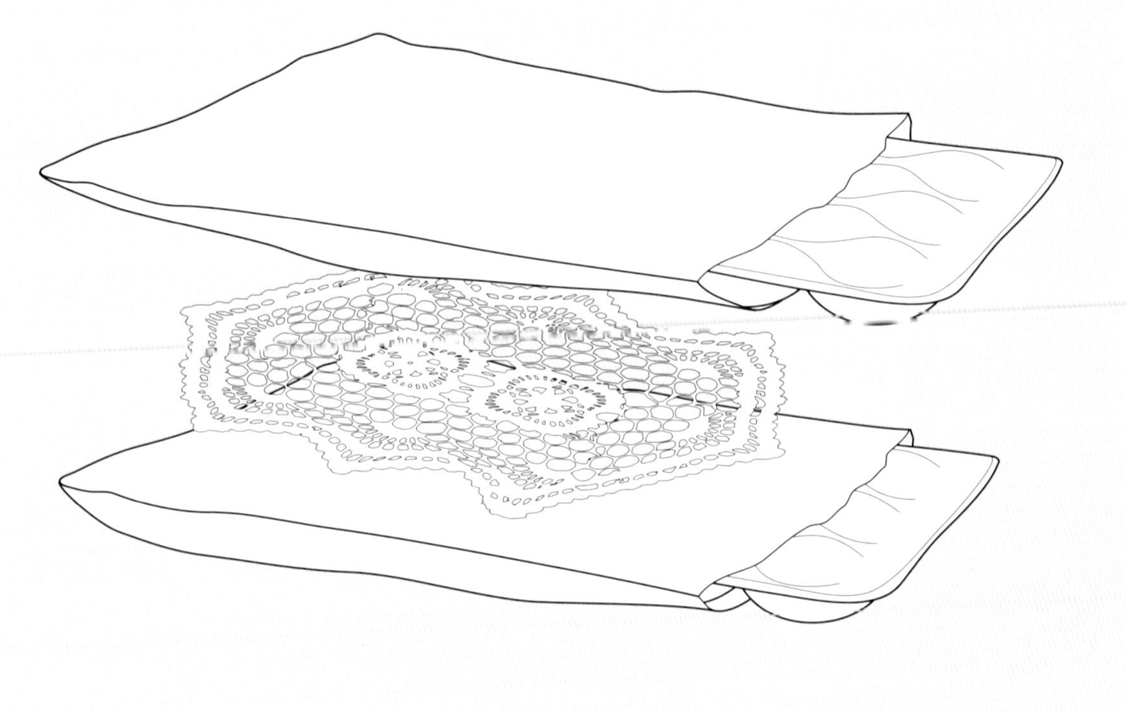

1 Trim your lace 3" larger on all sides than your anticipated fruit dish.

2 Lay a kitchen trash bag on a level surface for your work area.

3 Slip one of the cardboard fruit trays into a kitchen trash bag. This will be the bottom mold for your dish.

4 Slip the other cardboard fruit tray into another kitchen trash bag and set aside. This will be the top mold for your dish.

5 Lay your lace on top of the bottom mold.

6 Don the safety gloves. In the disposable mixing dish, mix the two elements of the epoxy compound in equal amounts and stir using a disposable plastic utensil.

7 Paint the lace with the compound in quick strokes. You want to saturate the material with as little clumping as possible. (You have twenty-five minutes of work time with the compound, but we suggest moving quickly.)

Number of $100 bills that can be made from a 480 bale of cotton: 313,600

THOUSAND THREAD-COUNT BATHMAT

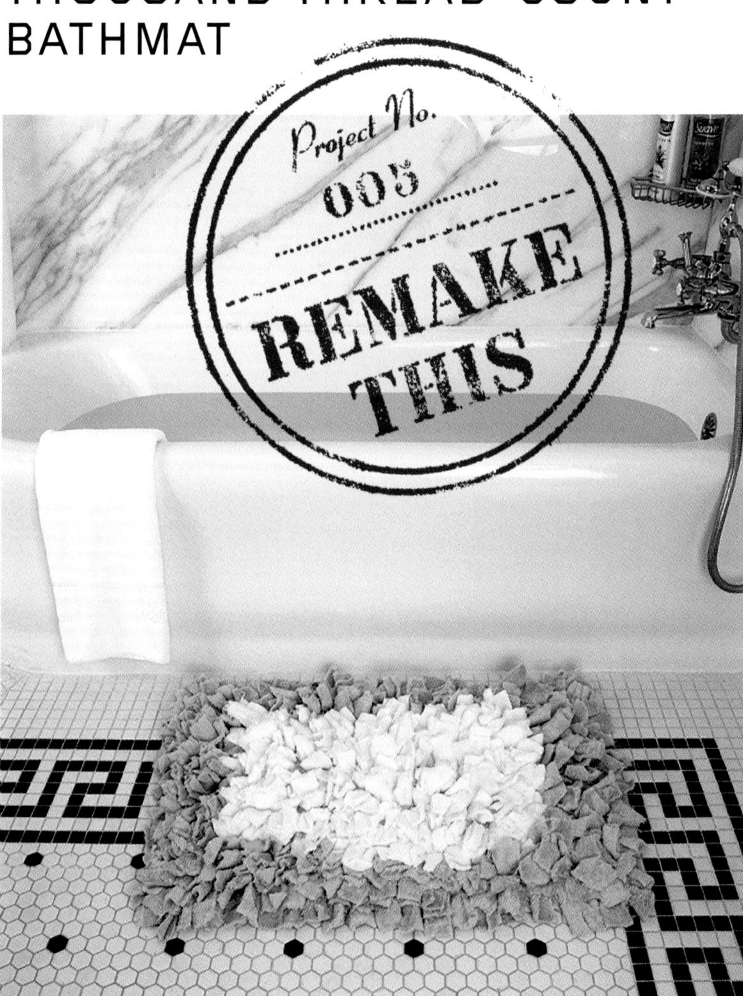

We thought we'd found another use for our frayed bathroom friends: Cut up old terry cloth into strips and hand-hook them through a bathmat-sized towel so the two ends poke out like tufts of grass. The resulting rug looks good, anyway.

YES

1 This foot-swaddler feels like a walk in the clouds.
2 It's cleverly made by latch-hooking into itself.
3 What else are you going to do with your old terry towels?

NO

1 If you don't use quality towels, you'll end up with terry-cloth crumbs everywhere.
2 Boy, was the process tedious.
3 You'll have to sew down each latch if you plan on washing the thing.

8 Flip the lace over and repeat step 7 on the other side.

9 When finished, set the top mold on top of the bottom, sandwiching the saturated lace in between.

10 Lay another plastic bag on top of the top mold.

11 Set a moderately heavy pile of magazines or books on top to ensure the rounded divots of the cardboard fruit dish are pressed together, without crushing the molds.

12 Let this sit undisturbed overnight. Because this is a chemical cure, it does not need air to dry.

13 Once dry, trim your molded lace with shears to size and break off any excess epoxy that may have clumped.

14 Pile on the fruit.

Meters of silk the average cocoon contains: 300–400

CARPET KIDDIE CHAIR

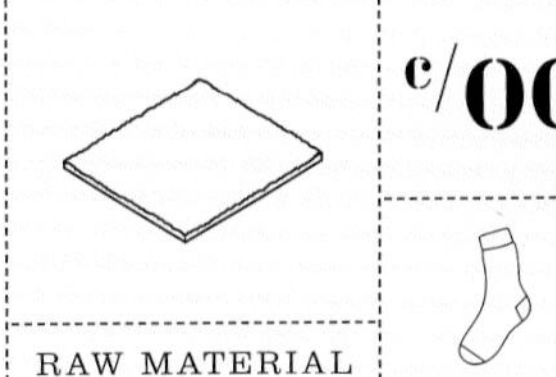

c/06

Problem: Kids are always banging themselves up on the edges of furniture. Solution: Remove all the edges! We used six square feet of soft carpet remnant, found at a local retailer, and a few rivets to make this cushy throne. When your enfant terrible opens up the salt shaker or pours the contents of his sippy-cup into the shag, it's easy to shake or vacuum clean. If only they made groovy furniture like this for grown-ups. Oh, wait, didn't Donna Summer have one of these in the '70s?

Number of silkworms it takes to produce 1 kg (2.2 lbs) of raw silk: **5,500**

CARPET KIDDIE CHAIR

$15

INGREDIENTS

+ 22" by 15" piece of carpet
+ 44" by 12" piece of carpet
+ Handful of 3/16" 5mm aluminum washers
+ Handful of 3/8" aluminum rivets

TOOLS

+ Carpet knife
+ Ruler
+ Rivet gun
+ Scissors
+ Blue masking tape
+ Awl

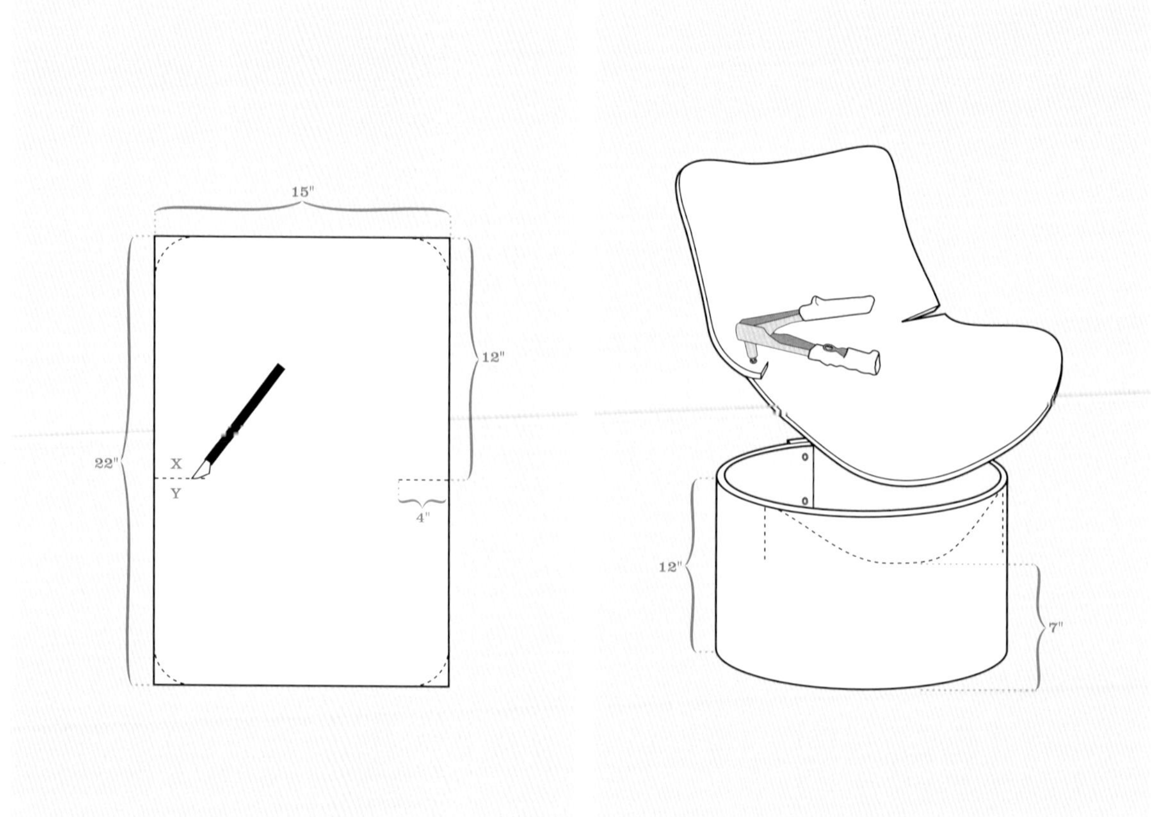

FIGURE A

FIGURE B

1 Lay out the 22" by 15" piece of carpet vertically on a flat work surface.

2 Using the carpet knife, cut two 4" horizontal slits 12" from the top of the carpet—one on each side. The taller portion will be the back of the chair, the shorter portion the seat (Figure A).

3 Now trim to round the corners.

4 Rivet corner X to corner Y. Do this on both sides for a concave seat shape (Figure B).

5 Take the 44" by 12" piece of carpet and form a tube with an approximately 2" overlap.

6 Rivet in three spots to secure the tube. You've just made the seat stem.

7 Set the seat shape on top of the sitting tube and measure 7" from the bottom. Mark a straight line with the blue tape.

8 Using your carpet knife, cut the

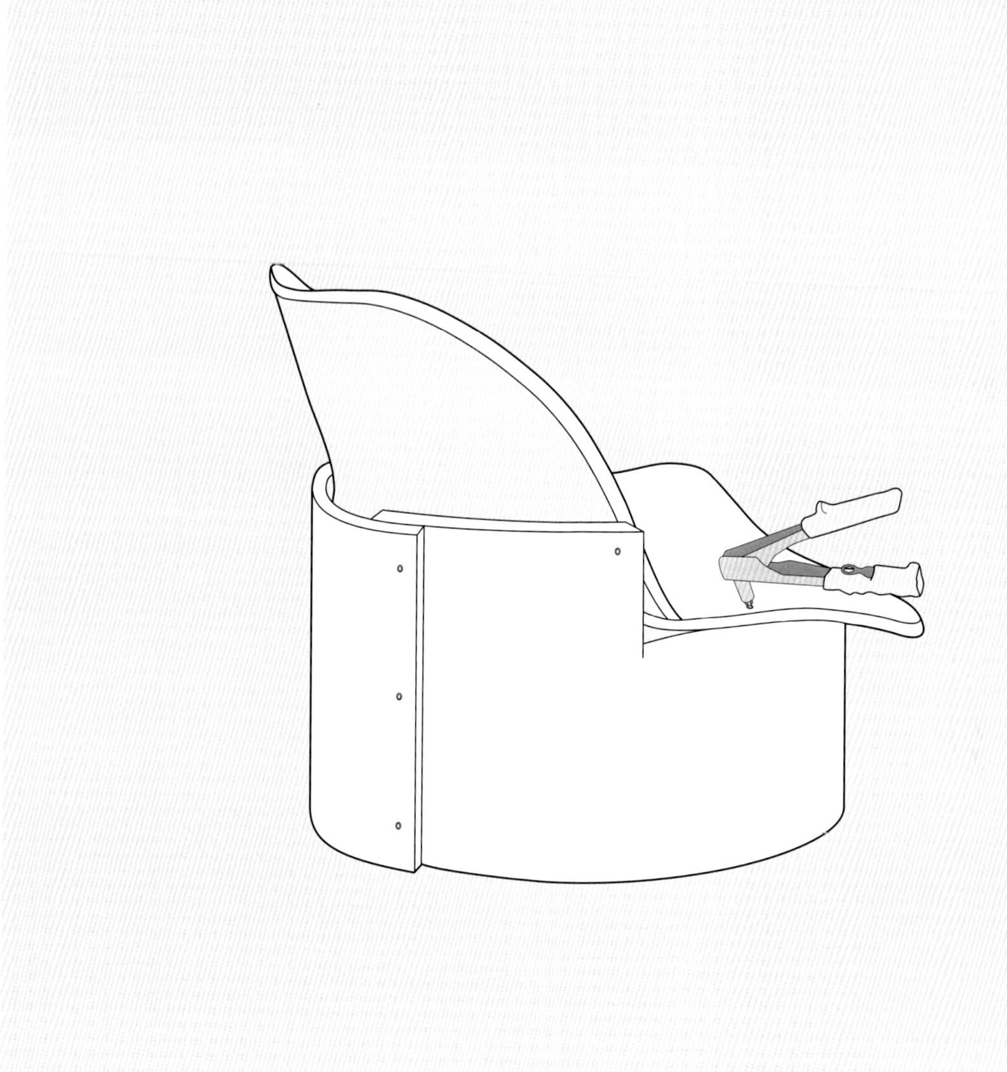

FIGURE C

tube where the backside of the seat touches the tube from end to end, dipping down to the 7" line. Again, this step can be performed freehand with appropriate tools (Figure B).

9 Cut the tube wall where the backside of the chair and the seat intersect (Figure B).

10 Rivet the backside of the seat to the chair stand (Figure C).

11 Rivet the folded part of the stem to the seat of the chair on both sides.

12 Stack some books or old magazines under the carpet seat for extra (hidden) structural support.

13 Place child on throne.

YOU, TOO, CAN BE A GREAT STORYTELLER, EVEN IF YOU WEREN'T BORN IN THE SOUTH

Spinning Yarn

How to Tell a Good Story

Once upon a time . . . Kidding! We're kidding. That's exactly what *not* to do—burden your story with platitudes, formulas, hackneyed devices. Many of you will be tempted to skip this part, having just learned how to deceive people. Are they not in the same phylum of fiction, you ask? They are not. Lying is to storytelling what table tennis is to Wimbledon. There is an art to yarn spinning that only people born in the South know instinctively. The skill can be learned later in life, but only with great patience and exertion. By way of example, let us tell you how the phrase "spin a yarn" came to mean regaling an audience.

The means by which any well-worn turn of phrase gains entry are anecdotal at best and indeterminate at worst, but we think it likely that women of the preindustrial era, while spinning wool into yarn at their jennies, passed the time by telling each other long, lavish stories of love and woe. The one about a poor cousin to whom an officer of some rank promised a better life on a Virginia plantation. (The leatherneck neglected to tell her that he had nine wives scattered throughout the Indies.) Or the one

There are certain instances where no other word but "penumbral" or "luminous" will do.

about the man who swallowed an electric eel and glowed like a streetlamp when in the presence of a woman he fancied. As the women worked their wool, the men were spinning their own yarn, in the form of tie-down ropes on ships. As legend has it, during wet weather, salty dogs would take shelter below deck and loosely twist together yarns of old twine, all the while telling each other tales of adventure and conquest. To ensure he gave the cord the proper torque, the spinner had to stretch the yarn out as he went—just as the truth is stretched a little to make a story more entertaining or surprising or dramatic. Like the one about the boy orphaned at sea when the ship carrying his parents capsized in a chaotic storm; how he learned to navigate and sailed the few remaining survivors to the nearest island just west of the Pillars of Hercules (now the Strait of Gibraltar), an empire populated by scantily clad, blue-eyed natives who claimed their empire to have been established by Poseidon, the god of the sea.

Telling a good story, you see, is like space travel. The beginning requires great propulsion—firing off petrochemical waves of heat and energy—to ignite the audience's interest. Then, when liftoff is established and you've got their chests pounding, you can deploy the main cargo, or thrust, of the story into orbit, where it will coast for awhile on very little fuel, and make time seem to slow down. Once you've got your audience in a flight path with you, it's time to bring things full circle with a fiery descent back into the atmosphere, finally concluding with a glorious landing or a ruinous crash (whichever suits).

That was what one calls an analogy. We dragged it out a little to make the point, but analogies are efficient devices that deliver their point by way of a correspondence between two sets of often unrelated things. I used a tennis analogy before to convey that telling a good tale requires much more skill than telling a simple lie. Use analogies when the point you're trying to make is difficult to grasp, or just to add a dash of color.

Use dramatic pauses and, on occasion, stop talking entirely. There's nothing like a good long break in the action to really set things up. Pauses are best delivered right before or after an important plot twist, cuing your audience that they may need to take a moment to settle in their chairs after what they've just heard. Add a generous sigh, or bite your lip and nod meaningfully, and you'll have them on the edge of their seats.

Use short, punchy words. Change *utilize* to *use*, *indicated* to *showed*, *however* to *but*. Don't waste long words on overly formal language; save them for descriptive passages where a flowery phrase is called for, like "her face was penumbral in the dusky light—dark, then luminous." There are certain instances where no other word but *penumbral* or *luminous* will do. A high-school English teacher taught us that we should all keep the three Cs in mind when writing a paper: clear, concise, and cogent. Keep it simple when you're relaying the bare bones of your plot, and use the lovelies sparingly.

Always use active verbs. You're not going to blow anyone's hair back with the passive voice. It's not that "she was seen by him in a bar with another man," it's that "he saw his mother at the bar, her hand on a strange man's knee!" Also, try to use the present tense whenever possible. Though this is a hotly contested point—(why would you use the present tense for something that occurred in the past?)—there's nothing like the present to carry an audience along and keep them engaged; it adds immediacy and drive.

Finally, never underestimate the element of surprise. Darth Vader is Luke's father? "Rosebud" was the sled? Bread is bad for you? Nothing gives a story more bang than pulling out a pistol no one saw coming. Why relate something that everyone knows the outcome of, anyway? If your story is short on surprises, manufacture them by adding obscure, digressive details about a person or place ("The restaurant where they met, it turned out, was once a brothel"), or upping the drama surrounding the central event with facial expressions—opening your eyes wide, shaking your head, or flashing some other can-you-believe-this sign.

c/06 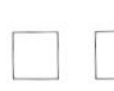 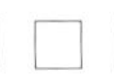 

Su Mo Tu We Th Fr Sa

Why Don't I Do This Every Day?

TESTIMONIAL # 0072

NAME: Akiko Ito

FROM: Tokyo, Japan

OCCUPATION: Designer, Crafter

Akiko, why don't you do these things every day?

Being a packrat, I make my life harder. But, look! There are so many things I can do with my accumulated junk. I just have to get up and start making instead of acting like a little animal preparing for a long winter.

1

MOVING PICNIC TOTE

INGREDIENTS

+ Moving blanket
+ 6' of webbing
+ 2 side-release buckles
+ 2 D-rings

TOOLS

+ Scissors
+ Sewing machine

01 Make your moving blanket into a square by cutting the longer side to match the shorter side in length.

02 Take the edging off the piece you trimmed and sew onto your freshly cut side.

03 Fold the square blanket in half diagonally, making an isosceles triangle.

04 Square off your top angles by folding in 4" and sewing it down. (Imagine folding the pointy flap of an envelope under so it is flat, not pointed.)

05 Fold up your right and left angles up to, but not past, the opposite side. The shape you have should look like a Chinese takeout box.

06 Fold this vertically into thirds so your squared-off corners look like the flap of a small envelope.

07 Pin and sew your webbing, buckles, and D-rings as seen in the photo.

08 Pack your lunch and have a picnic.

Why don't we?

2

PILLOW CASE POCKET

INGREDIENTS

+ Pillowcase
+ Magazine

TOOLS

+ Ruler
+ Sewing machine
+ Scissors

01 Turn your pillowcase inside out and reduce its width to 14" by sewing from end to opening and cutting off excess.

02 Turn right side out and fold the open end up 14" toward the closed end.

03 Cut the side seams of the pillowcase at the open end about 3" or to the length of the natural cuff.

04 Sew the folded 14" and the 3" flaps down to the side, creating two open pockets.

05 Cut open a 6" slit on the opposite closed end of the pillowcase on the right-hand side. This is where you will insert a small piece of cardboard or discarded magazine to hold your pocket in place between your mattresses.

06 Hook into your bed, insert your bedside goodies, and get those pajamas on.

We really should, you know.

3

TERRY APRON

INGREDIENTS

+ 3 hand towels
+ 12' of twill ribbon

TOOLS

+ Sewing machine
+ Scissors

01 Cut and arrange the towels in an apron shape, then sew them together to make a 13" by 17" panel.

02 Fold over both side edges 2" and sew together, leaving a finger's width of space to feed the twill ribbon through on both sides. You will end up with a finished panel of 13" by 13".

03 Cut and piece together the towels. Sew together to make a 21" by 20 $^{1}/_{2}$" panel.

04 On both of the upper corners of your larger panel, tuck in the corners 2" and sew down.

05 Center the smaller panel above the larger one and sew together.

06 Feed the twill ribbons through the sides of the smaller panel and through the tucked corners of the larger panel.

07 Put on your apron, wash your hands, and wipe them dry on your new creation.

Tarry no longer, chop chop!

APPENDIX

ABBREVIATIONS USED IN "THE PRACTICAL BOOK OF FORMULAS"

A.S.	Alcohol Soluble
AM.	Ammonium
AMP.	Ampere
APPROX	Approximately
ART.	Artificial
AV	Avoirdupois
BE.	Baume
C.	Centigrade
C.C.	Cubic Centimeter
C.D.	Current Density
C.P.	Chemically Pure
CARB.	Carbonate
COM	Commercial
COMPN.	Composition
CONCN.	Concentration
CONTG.	Containing
D.	Density
DEN	Denature
DIL.	Dilute
ESS.	Essence
EXT.	Extract
F.	Fahrenheit
F.F.C.	Free from Chlorine
F.F.P.A.	Free from Prussic Acid
FL	Fluid
GM. or G.	Gram
INF.	Infusion
KILO. KG.	Kilogram
L. LIT.	Liter
LB.	Pound
M.P.	Melting Point
NEUT.	Neutral
POT.	Potassium
POWD.	Powdered
PPTE.	Precipitate
PULV.	Pulverized
Q.S.	Quantity Sufficient
R.P.M.	Revolutions Per Minute
R.S.	Regular Soluble
S.D.	Semi Denatured
SAPON.	Saponified
SATD.	Saturated
SOD.	Sodium
SOLN.	Solution
SP. GR.	Specific Gravity
TINCT.	Tincture
U.S.P.	United States Pharmacopoeia
V	Voltage
VISC.	Viscosity

HARDWARE SCREW SIZES

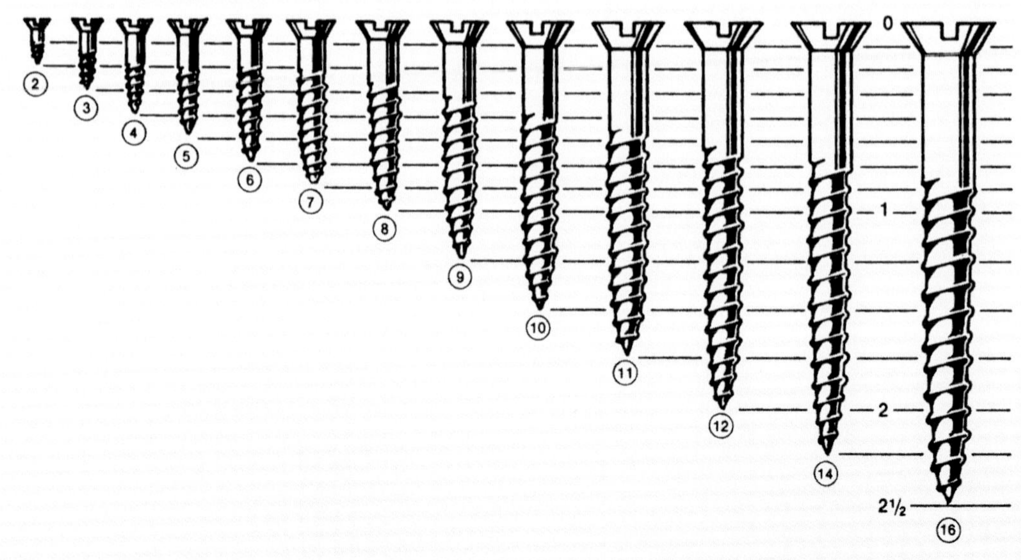

CONVERSIONS

1 centimeter (cm)	=	10 millimeters (mm)	=	0.3937 inches
1 decimeter (dm)	=	10 centimeters	=	3.937 inches
1 meter (m)	=	10 decimeters	=	3.3 feet
1 dekameter (dk)	=	10 meters	=	11 yards
1 hectometer (hm)	=	10 dekameters	=	110 yards
1 kilometer (km)	=	10 hectometers	=	1100 yards

INCHES = CENTIMETERS

1 in.	=	2.5 cm	36 in.	=	90 cm
6 in.	=	15 cm	40 in.	=	100 cm (1 m)
8 in.	=	20 cm	4 ft.	=	1.2 m
10 in.	=	25 cm	5 ft.	=	1.5 m
12 in.	=	30 cm	6 ft.	=	1.8 m
14 in.	=	35 cm	7 ft.	=	2.0 m

QUICK CONVERSIONS

knowing:		multiply by:		to obtain:
millimeters	x	0.04	→	inches
centimeters	x	0.40	→	inches
meters	x	3.30	→	feet
meters	x	1.10	→	yards
kilometer	x	0.60	→	miles
inches	x	2.50	→	centimeters
feet	x	30.00	→	centimeters
yards	x	0.90	→	meters
miles	x	1.60	→	kilometers
grams	x	0.035	→	ounces
kilograms	x	2.2	→	pounds
tons (1000 kg)	x	1.1	→	short tons

GLOSSARY

alloys: homogeneous mixture or solid solution of two or more metals.

awl: pointed tool for making holes in things.

bar clamp: tool with opposing sides that are adjusted by twisting up and down a long bar, for bracing objects or holding them together.

bubble wrap: used to protect fragile materials; also fun to pop.

carpenter's nails: long nails used for nailing into wood.

cable ferrule: metal ring placed around a cable to prevent splitting.

carpet tape: double-sided sticky stuff to hold rugs in place.

caster: small wheel on a swivel, attached under heavy furniture to make it mobile.

cellulose: complex carbohydrate found in trees that's composed of glucose. Used to manufacture much of our paper, textiles, and pharmaceuticals.

chopsticks: two slender sticks of wood, held between thumb and fingers as utensil; also a simple ditty played on the piano.

chlorine dioxide: ClO_2 used in bleaching paper, starch, soap, flour, and in water purification.

circular saw: power saw for cutting wood or metal consisting of a toothed disk rotated at high speed; also called buzz saw.

compound: created by combining two or more parts; to settle something (like a debt), or exacerbate it.

conduit strap: horseshoe-shaped bar with screw holes at each end to strap down a tube or duct for enclosing electric wires.

cordless drill: tool with pointed end used for boring holes and affixing screws.

contact paper: shelf-protecting paper or transluscent paper that gives the surface covering a diffuse, abstract look.

cut a rug: perform impressive dance moves.

diagonal pliers: hand tool having a pair of small pivoted jaws, used for holding, bending, or cutting.

dilute: to make thinner or less concentrated by adding a liquid (more tonic, less vodka).

D-ring: metal ring for hanging that is shaped like a capital D.

ductile: easily molded or shaped.

epoxy glue/compound: über-strong adhesive.

eyebolt: bolt with a looped head designed to attach hooks or ropes.

feathered quills: the shaft of a feather used as a writing instrument.

ferrous: containing iron.

fibers: long thread-like object or structure.

finish nails: small nails with a slight dent, or cup, in the head that are made to disappear in trim or other fine woodworking details.

flat-head Phillips screw: screw with a flat head and crossed slots.

foam pipe insulation: rigid polyurethane foam shaped like a tube with a slit, used for the insulation of plumbing pipes.

Galileo: Italian astronomer and mathematician who was the first to use a telescope to study the heavens.

galvanized floor flange: flat circular metal piece with screw holes that anchors plumbing pipe.

glassblowing: the process of shaping an object from molten glass or blowing air into it.

glue: sticky liquid; keeps everything in its place.

Goo Gone: miracle worker; combination of citrus and science that removes everything from label gum to tree sap to bicycle grease.

hacksaw: tough, fine-toothed blade used for cutting metal.

hammer: hand tool with handle attached perpendicularly to heavy metal head used for striking or pounding.

hand circular saw: power saw operated with one hand for making quick cuts.

hand drill: small power drill operated with one hand.

heating duct: large round metal pipe through which furnaces blow their heat.

industrial scissors: cutting implement consisting of two blades joined by a swivel pin allowing for edges to be opened and closed.

iron: metal appliance with weighted bottom that when heated is used to press wrinkles flat; mineral found in meat, fish, and greens that keeps blood healthy.

jigsaw: power-driven saw with narrow vertical blade, used to cut sharp curves.

keyhole screws: screws that fit into keyhole-shaped openings used to mount objects to the wall.

Ladderloc buckle: buckle that allows you to tighten a strap singlehandedly.

lignin: chief noncarbohydrate of wood; binds cellulose fibers, hardens and strengthens cell walls of plants.

lap link: chain link with an open, overlapping end used to repair broken chain.

linesman pliers: blunt-nosed grippers.

metal: elements that usually have a shiny surface, are good conductors of heat and electricity, and can be melted or fused, hammered into thin sheets, or drawn into wires; loud, thrashy music that evolved out of the blues in the early '70s in the UK.

metal shears: metal tool used to remove fleece or hair by cutting or clipping.

modge podge: water-based sealer, glue, and laminating finish for all surfaces.

monofilament twine: a lot like a fishing line.

mortar: building material made by mixing cement, sand, and water.

Mylar: transparent polyester film.

nonferrous: not composed of iron.

nose pliers: long, narrow-nosed grippers.

organic: derived from living organisms.

O-ring: rubber ring shaped like an O.

paintbrush: tool with horsehair or synthetic bristles on the end used to apply paint.

periodic table: arrangement of the elements according to their atomic number.

Phillips head: screw with a head having two intersecting perpendicular slots.

phobophobia: fear of phobias.

picture-hanging claw: toothed plate affixed to the back of a frame that hooks on a mounted nail.

pliers: hand tool used for holding, bending, or cutting.

Plexiglas: trademark material used for a light, transparent, weather-resistant thermoplastic.

plumbing couplings: plumbing device that links or connects.

plumbing hex bushing: plumbing screw.

plumbing nipples: threaded plumbing fitting with no head.

plywood: structural material made of multiple layers of wood glued together.

polyurethane: any of various resins used as a chemical- and water-resistant coating. Comes in gloss or matte.

rivet: headed pin or bolt of metal inserted through aligned holes in pieces to be joined, then hammered to make a second head.

rivet gun: hand-powered device that shoots the rivet into material.

roofing nails: a washer rests under the nailhead and the shank has a screw grip.

ruler: instrument used for drawing straight lines and measuring lengths.

sandpaper: heavy paper coated on one side with sand or other abrasive material, used for smoothing surfaces.

sawhorses: frame used to support pieces of wood being cut.

screw eye: a screw with a circular ring at the head.

Sharpie: everyone's favorite permanent marker; autograph tool of choice.

shelving angles: metal bracket with screw holes used to mount shelving.

side-release buckle: used on backpacks, bags, belts, and clothing; perhaps the most versatile buckle made.

silkworm: type of caterpillar that produces silk cocoons, which is the fiber source for commercial silk.

sodium carbonate: Na_2CO_3; white powdery compound used in manufacture of baking soda, glass, ceramics, detergents, and soap.

sodium sulfide: Na_2S; used as a metal ore reagent in photography and printing.

spade bit: thin drill bit with a center point and cutting edges on either side used to bore large holes.

steel-pipe nipples: tube threaded at both ends used as a plumbing fixture.

stud finder: locates the studs in walls so that you can mount heavy objects on them.

synthetic fiberfill: man-made pillow-stuffing that replaces the downier feathers of old.

thermoplastics: soft when heated, hard when cooled.

utility knife: a knife with a small, sharp, retractable blade.

Varathane (semi-gloss): a less toxic, water-based polyurethane finish that seals and coats wood, paper, and fabric.

washer: a flat disk, as of metal, plastic, rubber, or leather, placed beneath a nut or at an axle bearing or a joint to relieve friction, prevent leakage, or distribute pressure.

webbing: strong, narrow, closely woven fabric used for seatbelts and straps.

wood pallet: rigid platform used in shipping a variety of products to transport products efficiently with minimum damage to goods. About 40 percent of all hardwood harvested in the United States is for the construction of pallets.

wood veneer tape: a thin, plastic tape used for finishing the edges of wood; made to match many different wood varieties.

wood dowel: round wood pin that fits tightly into a corresponding hole to fasten or align two adjacent pieces; think clothing rod.

X-Acto knife: industry standard cutting tool for craft projects with a thin handle and extremely sharp, replacable blade that comes to a point at the end.

zoo: park or institution where living animals are kept and usually exhibited to the public.

CREDITS & ACKNOWLEDGMENTS

The following people made this book possible:

Wordsmithers: Clare Chatfield, Sean Cooper
Tinkers: Claire Bigbie, Pete Glover, Chelsa Robinson
Researchers: David Spataro, Emily Steen, Casey Tilmanis
Purveyors of photos and other fine imagery: David Byrne, Brian Holliday, Bob Willoughby, Walker Art Center
Designers: Adam Brodsley, Elizabeth Fitzgibbons, Akiko Ito
Stylist: Kristiana Lyons
Locations locations locations: Tina Barseghian, Adam Brodsley, James Chiang, Alex Lloyd, Marc Perlson
Secret agent man: Sloan Harris, ICM

SB dedicates this book to: The answer found in Tony Saxe
GH dedicates this book to: The boys in her life: Todd, Garrett, and George

Photo Credits:
David Byrne: p. 168; Corbis: pp. 34–35, 151–53 ; CSA Archive: p. 122; Getty Images: pp. 10, 44, 76, 108; Walker Art Center Archive: p. 1; Bob Willoughby: pp. 66–67.

Shoshana Berger came up with the idea for *ReadyMade* in 2000, just as the world was supposed to end. But the world did not end, and she became the magazine's editor in chief. She has written for the *New York Times, Wired, SPIN, Business 2.0,* and *Popular Science,* among others. She is currently attempting to write books.

Grace Hawthorne is the cofounder and CEO/Publisher of *ReadyMade* magazine. She has alternated between the worlds of business and art, earning both an MBA and MFA from UCLA, working as a strategic consultant in the entertainment/media industry, and serving as a board member for the Berkeley Art Museum, Humanities West, and SFMoMA's Media Arts Council. For her, *ReadyMade* represents the perfect marriage of art and business.

INDEX

Page numbers in *italics* refer to illustrations.

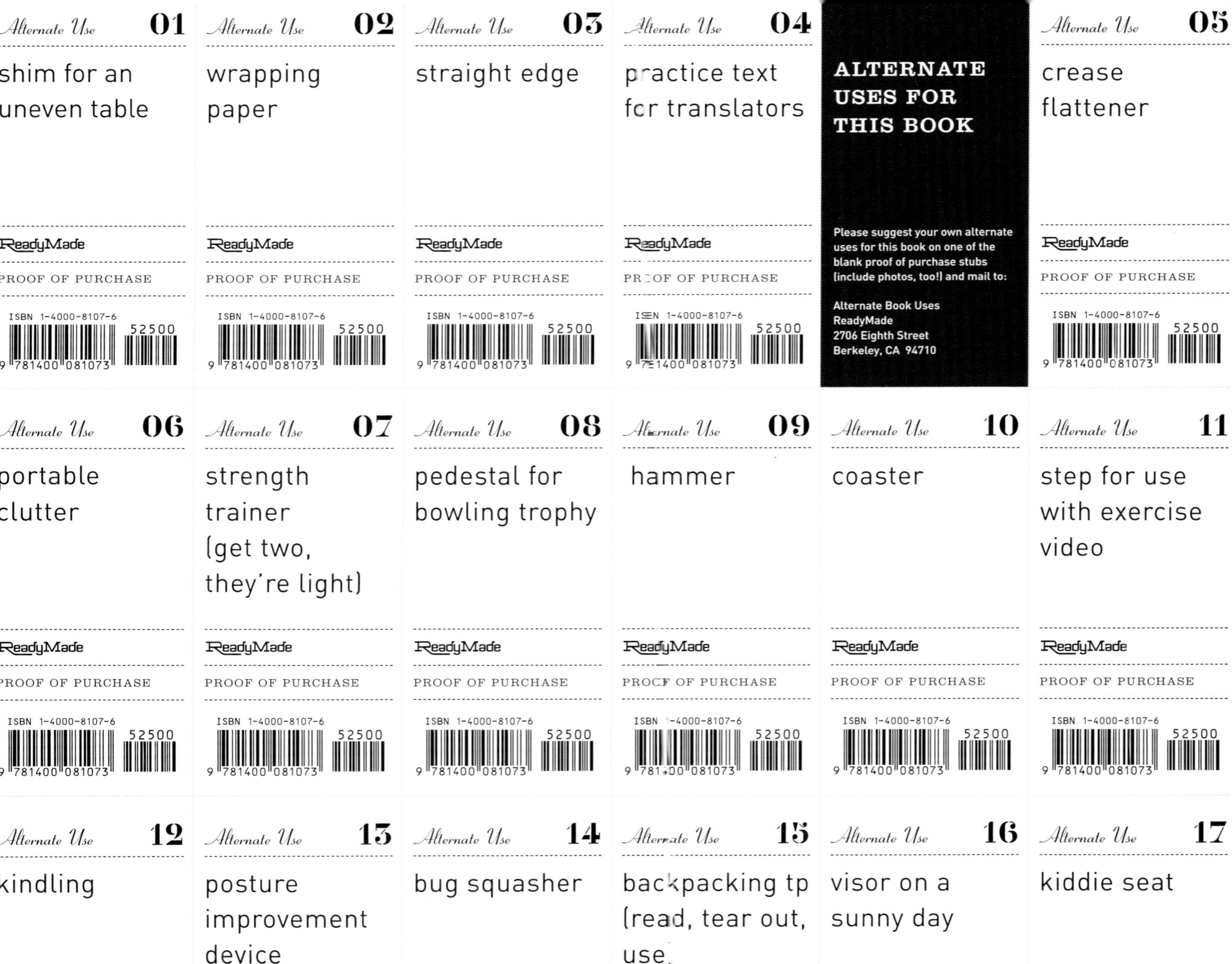

Alternate Use 01
shim for an uneven table
ReadyMade
PROOF OF PURCHASE
ISBN 1-4000-8107-6
9 781400 081073 52500
Alternate Use 02
wrapping paper
ReadyMade
PROOF OF PURCHASE
ISBN 1-4000-8107-6
9 781400 081073 52500
Alternate Use 03
straight edge
ReadyMade
PROOF OF PURCHASE
ISBN 1-4000-8107-6
9 781400 081073 52500
Alternate Use 04
practice text for translators
ReadyMade
PROOF OF PURCHASE
ISBN 1-4000-8107-6
9 781400 081073 52500
ALTERNATE USES FOR THIS BOOK
Please suggest your own alternate uses for this book on one of the blank proof of purchase stubs (include photos, too!) and mail to:
Alternate Book Uses
ReadyMade
2706 Eighth Street
Berkeley, CA 94710
Alternate Use 05
crease flattener
ReadyMade
PROOF OF PURCHASE
ISBN 1-4000-8107-6
9 781400 081073 52500
Alternate Use 06
portable clutter
ReadyMade
PROOF OF PURCHASE
ISBN 1-4000-8107-6
9 781400 081073 52500
Alternate Use 07
strength trainer (get two, they're light)
ReadyMade
PROOF OF PURCHASE
ISBN 1-4000-8107-6
9 781400 081073 52500
Alternate Use 08
pedestal for bowling trophy
ReadyMade
PROOF OF PURCHASE
ISBN 1-4000-8107-6
9 781400 081073 52500
Alternate Use 09
hammer
ReadyMade
PROOF OF PURCHASE
ISBN 1-4000-8107-6
9 781400 081073 52500
Alternate Use 10
coaster
ReadyMade
PROOF OF PURCHASE
ISBN 1-4000-8107-6
9 781400 081073 52500
Alternate Use 11
step for use with exercise video
ReadyMade
PROOF OF PURCHASE
ISBN 1-4000-8107-6
9 781400 081073 52500
Alternate Use 12
kindling
Alternate Use 13
posture improvement device
Alternate Use 14
bug squasher
Alternate Use 15
backpacking tp (read, tear out, use.
Alternate Use 16
visor on a sunny day
Alternate Use 17
kiddie seat

ReadyMade
PROOF OF PURCHASE
ISBN 1-4000-8107-6
9 781400 081073 52500
Alternate Use 18
stocking stuffer
ReadyMade
PROOF OF PURCHASE
ISBN 1-4000-8107-6
9 781400 081073 52500
Alternate Use 19
bulletproof shield
ReadyMade
PROOF OF PURCHASE
ISBN 1-4000-8107-6
9 781400 081073 52500
Alternate Use 20
flower press
ReadyMade
PROOF OF PURCHASE
ISBN 1-4000-8107-6
9 781400 081073 52500
Alternate Use 21
TV stand
ReadyMade
PROOF OF PURCHASE
ISBN 1-4000-8107-6
9 781400 081073 52500
Alternate Use 22
helium balloon holder
ReadyMade
PROOF OF PURCHASE
ISBN 1-4000-8107-6
9 781400 081073 52500
Alternate Use 23
book end
ReadyMade
PROOF OF PURCHASE
ISBN 1-4000-8107-6
9 781400 081073 52500
Alternate Use 24
(write your alternative use here)
ReadyMade
PROOF OF PURCHASE
ISBN 1-4000-8107-6
9 781400 081073 52500
Alternate Use 25
(write your alternative use here)
ReadyMade
PROOF OF PURCHASE
ISBN 1-4000-8107-6
9 781400 081073 52500
Alternate Use 26
(write your alternative use here)
ReadyMade
PROOF OF PURCHASE
ISBN 1-4000-8107-6
9 781400 081073 52500
Alternate Use 27
(write your alternative use here)
ReadyMade
PROOF OF PURCHASE
ISBN 1-4000-8107-6
9 781400 081073 52500
Alternate Use 28
(write your alternative use here)
ReadyMade
PROOF OF PURCHASE
ISBN 1-4000-8107-6
9 781400 081073 52500
Alternate Use 29
(write your alternative use here)
ReadyMade
PROOF OF PURCHASE
ISBN 1-4000-8107-6
9 781400 081073 52500

First published in the United Kingdom in 2006 by
Thames & Hudson Ltd, 181A High Holborn, London WC1V 7QX

www.thamesandhudson.com

British Library Cataloguing-in-Publication Data
A catalogue record for this book is available from the British Library

ISBN-13: 978-0-500-51338-5
ISBN-10: 0-500-51338-4

Design by Volume Design, Inc.
Printed and bound in Singapore by Star Standard Industries (Pte) Ltd

WRONG